谨以此书纪念北京邮电大学
国际学院建院二十周年
（2004—2024年）

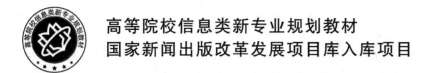

高等院校信息类新专业规划教材
国家新闻出版改革发展项目库入库项目

物联网无线技术应用场景实践
（中英双语版）

Practice of IoT Wireless Technology in Application Scenarios
(Chinese-English Edition)

主编 郭 凯
参编 段鹏瑞 鲍 东 姚东伟 边立涛
　　 陈云峰 史琳娟 孙 晔

北京邮电大学出版社
www.buptpress.com

内 容 简 介

物联网作为信息技术与实体世界深度融合的重要领域,已成为推动经济发展、提升社会效益的重要力量。本教材从应用场景的角度出发,全面讲述了物联网的发展情况、主流的无线接入技术。在不同的应用场景下,本教材结合各场景的应用特点,从分析需求、选择技术和产品、搭建系统到开通业务的各环节,对各个实践项目流程进行了详述,力求能为行业相关人员解决部分疑惑和问题。

本教材适合所有对物联网感兴趣的人阅读,无论读者是初学者还是物联网从业者,都可以从本教材中受益。

图书在版编目(CIP)数据

物联网无线技术应用场景实践:汉、英 / 郭凯主编.
北京:北京邮电大学出版社,2025. -- ISBN 978-7-5635-7503-9

Ⅰ. TP393.4;TP18;TN92

中国国家版本馆 CIP 数据核字第 2025RP2038 号

策划编辑:刘纳新　责任编辑:王晓丹　杨玉瑶　责任校对:张会良　封面设计:七星博纳

出版发行:北京邮电大学出版社
社　　　址:北京市海淀区西土城路 10 号
邮政编码:100876
发 行 部:电话:010-62282185　传真:010-62283578
E-mail:publish@bupt.edu.cn
经　　　销:各地新华书店
印　　　刷:保定市中画美凯印刷有限公司
开　　　本:787 mm×1 092 mm　1/16
印　　　张:11.5
字　　　数:257 千字
版　　　次:2025 年 3 月第 1 版
印　　　次:2025 年 3 月第 1 次印刷

ISBN 978-7-5635-7503-9　　　　　　　　　　　　　　　　　　　　　　　　定价:38.00 元

·如有印装质量问题,请与北京邮电大学出版社发行部联系·

前　言

随着科技的不断发展,物联网已经成为我们生活和工作中不可或缺的一部分,物联网技术正逐渐渗透到我们生活的方方面面。它如同一个神奇的"网间网",将各种实体相互连接,从家中的智能家电到工厂的自动化生产线,从农田的智能灌溉系统到城市的交通信号灯,甚至是地球上每一个角落的生命体。物联网正在重塑我们的世界,为我们带来了前所未有的便利和可能性。

本教材正是为了帮助读者更好地理解和掌握物联网技术而编写的。我们将深入探讨物联网的核心概念、技术原理、应用场景以及未来发展趋势,通过丰富的案例和场景分析,让读者全面了解物联网的魅力和潜力。

在本教材中,读者将学习物联网的技术特点,包括无线接入技术、常见应用场景,以及部署物联网项目的相关步骤和注意事项等方面的知识。同时,我们还将通过实际案例,介绍物联网在智能家居、智慧农业、智能运输等领域的应用,使读者能够深入了解物联网在不同领域的应用和优势,以及如何为企业和政府带来创新和价值。

我们希望通过本教材的介绍和案例分析,能够激发读者对物联网的兴趣和热情。同时,我们也希望本教材能够成为读者在物联网领域探索和实践的指南,帮助读者在这个充满机遇和挑战的领域取得成功。

本教材是由北京邮电大学国际学院和北京中兴协力科技有限公司合作完成的,两单位在本科教育和工程实习方面有着长期且紧密的合作。校企合作撰写教材有着无可比拟的优势,主要包括以下 6 点:

(1) 紧密结合行业需求:北京中兴协力科技有限公司对于通信、物联网行业的最新动态和市场需求有深入的了解,通过与学校合作编写教材,可以将这些实际需求融入教材,使教材内容更加贴近实际,更具实用性。

(2) 引入实际案例:北京中兴协力科技有限公司长期致力于与运营商、政企相关的业内专业培训,在与行业专家交流接触的过程中,可以参与了解大量的项目,为本教材提供丰富的实际案例,这些案例可以帮助读者更好地理解理论知识,并提高其在实际工作中的应用能力。

(3)促进学生就业:校企合作编写的教材可以更好地满足企业的用人需求,提高学生的就业竞争力,同时,企业也可以通过教材了解学生的实际情况,为其招聘提供更加准确的参考。

(4)实现资源共享:学校和企业各自拥有不同的资源和优势,通过合作编写教材,可以实现资源共享,提高资源的利用效率。

(5)推动教学改革:校企合作编写教材可以推动学校的教学改革,使教学内容更加符合实际需求,优化教学效果。

(6)促进产学研结合:校企合作编写教材有助于促进产学研结合,推动科学研究和技术创新的发展,为企业和学校的发展提供技术支持和人才储备。

总之,校企合作编写教材可以实现学校与企业的优势互补,提高教材质量和实用性,促进学生就业和教学改革,推动产学研结合和企业发展。

最后,我们要感谢所有为本教材做出贡献的人员,特别是那些在物联网领域中辛勤工作、不断创新的专家和工程师们。同时,我们也希望本教材能够对读者的学习和工作有所帮助,成为其迈向物联网未来的有力助手。

祝您学习愉快!

作　者

目 录

第 1 章 物联网概述 1

1.1 全球物联网市场情况 1
1.2 中国物联网发展情况 2
1.3 物联网的主要技术 3

第 2 章 物联网无线技术综合概述 7

2.1 短距通信 7
    2.1.1 Bluetooth 7
    2.1.2 Wi-Fi 13
    2.1.3 ZigBee 24
2.2 低功耗远距通信 28
    2.2.1 Sigfox 28
    2.2.2 NB-IoT Technology 34
    2.2.3 LoRaWAN Technology 41
2.3 私有协议 47
    2.3.1 华为星闪技术 47
    2.3.2 CLAA 52
    2.3.3 Mesh Network 57

第 3 章 物联网场景实践 67

3.1 智慧社区 67
    3.1.1 Smart Home 67

3.1.2　Smart Community……………………………………………………………… 74
　　3.1.3　智慧养老…………………………………………………………………………… 91
3.2　绿色双碳……………………………………………………………………………………… 97
　　3.2.1　算力中心环境监测………………………………………………………………… 97
　　3.2.2　Building Safety……………………………………………………………… 102
3.3　乡村振兴……………………………………………………………………………………… 131
　　3.3.1　智慧仓储…………………………………………………………………………… 131
　　3.3.2　Digital Logistics…………………………………………………………… 138
3.4　商业数字化升级领域——智慧零售……………………………………………………… 146
　　3.4.1　行业背景…………………………………………………………………………… 146
　　3.4.2　项目方案…………………………………………………………………………… 148
　　3.4.3　合作案例…………………………………………………………………………… 150
3.5　创新创业实践应用案例…………………………………………………………………… 150
　　3.5.1　案例1：基于多源融合感知的智能博物馆消防应急系统…………………… 151
　　3.5.2　案例2：智能会展系统…………………………………………………………… 157
　　3.5.3　案例3：基于物联网的智慧停车装置和线上平台……………………………… 161

参考文献………………………………………………………………………………………… 169

附录　LoRaWAN 实验箱说明………………………………………………………………… 170

第1章 物联网概述

1.1 全球物联网市场情况

近年来,全球物联网市场规模呈现出强劲的增长势头,市场规模、终端连接数量预计在未来几年保持高速增长。数据显示,2020年全球物联网市场规模达到了约1.36万亿美元,且近两年的复合年增长率稳定在10%以上。而在接下来的几年中,这一数字还将持续增长。预计到2026年,全球物联网市场规模将达到6505亿美元。更为乐观的是,从2022年到2027年,预测全球物联网市场规模将以19.4%的复合年增长率增长,并在2027年达到1.2万亿美元。

各国高度重视物联网的发展,纷纷从战略高度制定物联网策略,以抢占发展先机。美国利用其在芯片、软件、互联网等领域的技术优势,大力推进物联网在军事、电力、工业等多个领域的应用。美国联邦通信委员会(FCC)发起了"US Cyber Trust Mark"网络安全标签计划,该计划旨在帮助消费者筛选出安全的物联网设备,保护消费者权益。美国网络安全和基础设施安全局(CISA)在2024财年预算中获得了31亿美元的拨款,创历史新高,该笔拨款将用于提高网络安全和分析能力,其中也涉及物联网安全的投入。欧盟先后组建了物联网创新联盟(AIOTI)、物联网创新平台(IoT-EPI),致力于构建可持续发展的物联网生态系统。日本联合2 000多家国际企业组成了物联网推进联盟,其国内物联网市场规模预计将从2016年的6万亿日元,增长到2020年的14万亿日元。韩国以人工智能、物联网城市等九大国家创新项目作为拉动国民经济增长的新动力,积极规模化部署物联网商用网络。俄罗斯在物联网技术发展路线草案的基础上拟定超过20个物联网试点项目。

特别是在亚太地区,物联网市场规模的增长潜力尤为显著。预测在2022年至2027年期间,亚太地区的物联网市场规模将以22%的复合年增长率增长,并有望超过世界其他地区。

赛迪的数据显示,物联网的传输层占据最大份额,但随着各领域市场需求的释放,平台层和应用层市场的增长速度将持续呈上升趋势。未来几年,物联网应用将从闭环、碎片化走向开放、规模化,智慧城市、工业物联网、车联网等领域将率先突破。IDC进一步预测,到2025年,全球物联网市场规模有望达到1.2万亿美元,复合年增长率达到11.4%。

总的来说,全球物联网市场不仅规模巨大,而且仍在高速增长中,这预示着未来有更多的机会和挑战等待着各行业和企业。

1.2　中国物联网发展情况

我国政府高度重视物联网产业的发展。早在2010年,国务院出台了《国务院关于加快培育和发展战略性新兴产业的决定》,物联网作为新一代信息技术产业中的重要项目位列其中,成为国家首批加快培育的七个战略性新兴产业之一,标志着物联网的发展在中国已经上升为国家战略。随后我国又陆续出台了《物联网"十二五"发展规划》(2012)、《国务院关于推进物联网有序健康发展的指导意见》(2013)、《关于印发10个物联网发展专项行动计划的通知》(2013)、《工业和信息化部2014年物联网工作要点》(2014)、《关于开展2015年智能制造试点示范专项行动的通知》(2015)、《信息通信行业发展规划物联网分册(2016—2020年)》(2016)、《物联网新型基础设施建设三年行动计划(2021—2023年)》(2021)等,以保障我国物联网产业健康有序的发展。

经过二十多年的发展,目前,我国物联网产业规模庞大。据中国产业发展研究院公布的数据,2022年我国物联网市场规模约为3.05万亿元,同比增长15.97%;2023年我国物联网市场规模进一步增长至约3.5万亿元;预测到2024年,这一数字将增长至约4.31万亿元。

在物联网设备连接数方面,中国也取得了显著的增长。根据GSMA发布的 *The Mobile Economy* 2020报告,2019年全球物联网总连接数为120亿,而我国的物联网连接数占全球的30%,达到了36.3亿。到了2020年末,我国物联网的终端连接数已经达到45.3亿,预计到2025年能够超过80亿。

我国物联网应用发展进入实质性推进阶段。物联网的理念和相关技术产品已经广泛渗透到社会、经济及民生的各个领域,物联网凭借与新一代信息技术的深度集成和综合应用,在推动转型升级、提升社会服务、改善服务民生、推动增效节能等方面正发挥重要的作用,在部分领域正带来真正的"智慧"效应。国家智能电网管理物联网应用示范工程已进入全面实施阶段;不停车电子收费、路网监测、车辆管理和调度等交通领域应用开始发挥积极作用。我国在2015年底前已基本实现全国所有省区市的系统联网运行,用户数量达2 000万,主线收费站实现全覆盖;截至2014年底,中国在先后两批190余个城市进入智慧城市试点之后,迅速在年内增长至286个城市,实际投入运作的智慧城市试点项目达409个,在"十二五"期间,各地智慧城市建设带来了2万亿元的产业机会。当前,物联网应用正从政府逐步向企业市场、民生市场扩张,应用层次正不断得到深化。一些商业模式逐渐成熟,比如在智能交通、安全生产、医疗健康领域,人们已摸索出一套"政府公共服务+市场增值服务"的可持续运营模式。

物联网产业体系已经初步形成,我国已经形成涵盖MEMS加工和设计、网络通信、软件与信息处理、行业与社会服务等相对完整的物联网产业体系。产业规模不断扩大,我国已经形成环渤海、长三角、珠三角以及中西部地区4大区域集聚发展的空间布局,呈现出高端要素集聚发展的态势。其中,长三角地区是中国物联网技术和应用的起源地,产业规模较大;珠三角地区是国内电子整机的重要生产基地,具有物联网产业规模化发展的巨大潜能;环渤海地区是国内物联网产业重要的研发、设计、设备制造及系统集成基地;中西部地区依托在RFID、芯片设计、传感传动、自动控制等领域较好的产业基础,已形成了一定的产业特色。基于完整的产业链和空间布局,我国物联网产业近年来一直保持较高的速度增长。

1.3 物联网的主要技术

物联网是以感知为基础的物物互联系统,涉及网络、通信、信息处理、传感器、RFID、安全、服务、标识、定位、同步、数据挖掘、多网融合等众多技术领域。从技术角度讲,物联网涉及的专业有:计算机科学与工程、电子与电气工程、电子信息与通信、自动控制、遥感与遥测、精密仪器、电子商务等。物联网自上而下划分,其技术架构分为感知层、网络层和应用层3个层面。而物联网具有与其他学科交叉的特性,除3个层面涉及的关键技术外,还包括共性技术和支撑技术。因此,为了系统分析物联网技术体系,将物联网技术体系划分为感知层关键技术、网络层关键技术、应用层关键技术、共性技术和支撑技术五大类别,如图1.3.1所示。

1. 感知层关键技术

感知层由数据采集子层、短距离通信技术和协同信息处理子层组成。

① 数据采集子层通过各种类型的传感器获取物理世界中发生的物理事件和数据信息,如各种物理量、标识、音视频多媒体数据。物联网的数据采集涉及传感器、RFID、多媒体信息采集、二维条码和实时定位等技术。

② 短距离通信技术和协同信息处理子层将采集到的数据在局部范围内进行协同信息处理,以提高信息的精度,降低信息冗余度,并通过具有自组织能力的短距离传感器网络接入广域承载网络。感知层中间件技术旨在解决感知层数据与多种应用平台间的兼容性问题,包括代码管理、服务管理、状态管理、设备管理、时间同步、定位等。在有些应用中,感知层中间件技术还需要通过执行器或其他智能终端对感知结果做出反应,以实现智能控制。

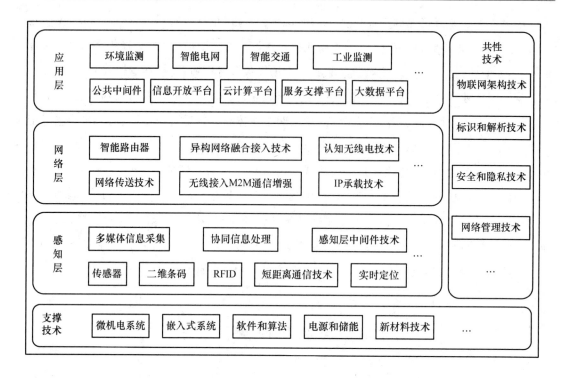

图1.3.1 物联网技术体系

2．网络层关键技术

网络层将来自感知层的各类信息通过基础承载网络传输到应用层，包括移动通信网、互联网、卫星网、广电网、行业专网及形成的融合网络等。网络通信关键技术主要实现物联网数据信息和控制信息的双向传递、路由和控制，重点包括智能路由器、自组织通信、无线接入 M2M 通信增强、IP 承载技术、网络传送技术、异构网络融合接入技术及认知无线电技术等。

3．应用层关键技术

应用层实现物联网信息资源的利用，根据底层采集的数据，形成与业务需求相适应、实时更新的动态数据资源库。综合运用高性能计算、人工智能、数据库、模糊计算等技术，对收集的感知数据进行通用处理，重点涉及海量存储、分布数据处理、数据挖掘、信息管理、平台服务、面向服务的体系架构（SOA）等。面向服务的体系架构是一种松耦合的软件组件技术，它将应用程序的不同功能模块化，并将其通过标准化接口和调用方式联系起来，实现快速可重用的系统开发和部署。SOA 可提高物联网架构的扩展性，提升应用开发效率，充分整合和复用信息资源。

4．共性技术

物联网共性技术涉及不同层面，主要包括物联网架构技术、标识和解析技术、安全和隐私技术、网络管理技术等。

(1) 物联网架构技术

物联网架构技术是从顶层设计角度清晰地描述物联网层级结构的技术,从总体上指导物联网通用体系架构各个组成部分及其交互关系,进而使各组成部分相互有机联系、形成整体,提高物体间的互联互通和互操作性,涉及顶层架构、系统接口、标准规范、协议及语言、安全设计等方面。

(2) 标识和解析技术

标识技术是对物品进行有效的、标准化的技术手段,能为物品分配唯一的标识符。解析技术是将标识符转换为可访问的信息和服务的过程。物联网标识和解析技术涉及不同的标识体系、不同体系的互操作、全球解析或区域解析、标识管理等。

(3) 安全和隐私技术

安全和隐私技术包括安全体系架构、网络安全技术、"智能物体"的广泛部署对社会带来的安全威胁、隐私保护技术、安全管理机制和保证措施等。

(4) 网络管理技术

网络管理技术重点包括管理需求、管理模型、管理功能、管理协议等。为实现对物联网广泛部署的"智能物体"的管理,人们需要进行网络功能和适用性分析,开发适合的管理协议。

5. 支撑技术

物联网支撑技术有很多,当前主要包括微机电系统(MEMS)、嵌入式系统、软件和算法、电源和储能、新材料技术等。随着物联网产业与其他产业的融合发展,其支撑技术范围也将不断扩展。

(1) 微机电系统

微机电系统可实现对传感器、执行器、处理器、通信模块、电源系统等的高度集成,是支撑传感器节点微型化、智能化的重要技术。

(2) 嵌入式系统

嵌入式系统是满足物联网对设备功能、可靠性、成本、体积、功耗等的综合要求,可以按照不同应用定制裁剪的嵌入式计算机技术,是实现物体智能的重要基础。

(3) 软件和算法

软件和算法是实现物联网功能、决定物联网行为的主要技术,重点包括各种物联网计算系统的感知信息处理、交互与优化软件与算法、物联网计算系统体系结构与软件平台研发等。

(4) 电源和储能

电源和储能是物联网关键支撑技术之一,包括电池技术、能量储存、能量捕获、恶劣情况下的发电、能量循环、新能源等技术。

（5）新材料技术

新材料技术主要是指应用传感器敏感材料实现的技术。传感器敏感材料包括湿敏材料、气敏材料、热敏材料、压敏材料、光敏材料等。新敏感材料的应用可以使传感器的灵敏度、尺寸、精度、稳定性等特性获得改善。

第 2 章 物联网无线技术综合概述

2.1 短距通信

2.1.1 Bluetooth

1. Development Background and Current Situation

Bluetooth technology is an advanced open wireless communication standard that can wirelessly connect desktop computers, laptops, portable devices, PDAs, mobile phones, camera phones, printers, digital cameras, headphones, keyboards, and mice within a short-distance range. Using Bluetooth technology, communication between mobile communication terminal devices can be effectively simplified, as well as communication between devices and the Internet, making data transmission faster and more efficient.

Bluetooth adopts a distributed network structure, fast frequency-hopping, and short-packet technology. It supports point-to-point and point-to-multipoint communication and operates in the globally available 2.4 GHz ISM (Industrial, Scientific, Medical) frequency band. It uses a time-division duplex transmission scheme to achieve full-duplex transmission. In short, Bluetooth technology enables wireless communication between various digital devices. With Bluetooth wireless technology, it is easy to connect computers and portable devices, mobile phones, and other peripheral devices wirelessly within a distance of 9 m.

Compared to other wireless technologies, such as infrared, wireless 2.4 GHz, and Wi-Fi,

Bluetooth has many advantages, including improved encryption measures, stable transmission processes, and compatibility with a wide range of devices. Especially today, when the authorization threshold is gradually decreasing, Bluetooth technology is truly becoming popular in all digital devices.

Currently, the latest protocol for Bluetooth technology is the Bluetooth Special Interest Group, which released the new-generation Bluetooth standard, Bluetooth 5, on June 16, 2016. Bluetooth 5 has a faster transmission speed and longer transmission distance compared to its predecessor.

Currently, the latest version of Bluetooth, 5.0, has greater technological advantages compared to previous versions, which are reflected in the following aspects.

① Faster transmission speed: The developers of Bluetooth 5.0 claim that the maximum transmission speed of the new version is 24 Mbit/s, which is twice that of the previous 4.2LE version.

② Longer effective range: Another important improvement of Bluetooth 5.0 is that its effective range is four times that of the previous version. In theory, the effective working distance between Bluetooth transmitting and receiving devices can reach 300 m.

③ Navigation function: Bluetooth 5.0 will add more navigation functions, so this technology can be used as an indoor navigation beacon or similar positioning device. When combined with Wi-Fi, it can achieve indoor positioning with an accuracy of less than 1 m. For example, if you are as directionally challenged as the translator, you can use Bluetooth technology to find your way in a large commercial center.

④ Internet of Things Functionality: The Internet of Things is still booming, so Bluetooth 5.0 has made many underlying optimizations for the Internet of Things, aiming to provide smarter home services with lower power consumption and higher performance.

⑤ Hardware Upgrade: Previous updates of some Bluetooth versions only require software upgrades, but Bluetooth 5.0 may require upgrading to a new chip. However, although old hardware is still compatible with Bluetooth 5.0, it cannot enjoy its new features. Flagship smartphones equipped with Bluetooth 5.0 chips have been released, and I believe that mid-range and low-end smartphones will also gradually incorporate Bluetooth 5.0 chips.

2. Technical Principles

The technical principle of Bluetooth is mainly based on the transmission of data through radio waves. It operates in the 2.4 GHz frequency band and uses Frequency Hopping Spread Spectrum (FHSS) technology for data transmission, with a theoretical hopping rate of 1 600 hop/s. Frequency-hopping technology divides the frequency band into several hopping channels. In a connection, the wireless transceiver continuously hops from one channel to another according to a certain code sequence (pseudo-random code). Only communicating parties follow this pattern, while other interferences cannot interfere in the same way. The instantaneous bandwidth of frequency-hopping is very narrow, but the spread spectrum technology expands this narrow frequency-band into a wide bandwidth, minimizing the potential impact of interference. With a center frequency of 2.45 GHz, a maximum of 79 channels with a bandwidth of 1 MHz can be obtained. In Japan, Spain, and France, the bandwidth of the frequency band is small, accommodating only 23 hopping points, with a 1 MHz interval. The Bluetooth channels are divided into time slots with a length of 625 μs, and the slots are numbered based on the Bluetooth clock of the micro-network. Bluetooth technology adopts a master-slave mode, where a master device (such as a mobile phone) can control multiple slave devices (such as Bluetooth headsets, and smart bands), and the slave devices transmit data by establishing a connection with the master device. In the Bluetooth system, the grouping transmission of the master and slave units adopts the Time Division Duplexing (TDD) alternating transmission method. The main unit starts information transmission in even-numbered slots, while the slave unit starts information transmission in odd-numbered slots. The starting position of the grouping coincides with the starting point of the slot, and the grouping transmitted by the master or slave unit can extend to 5 slots. Bluetooth adopts GFSK modulation and uses 3 power levels: 0 dBm (1 mW), 4 dBm (2.5 mW)、20 dBm (100 mW).

Between the master and slave units, different types of links can be established, such as synchronous connection-oriented links (SCO) and asynchronous connectionless links (ACL). SCO links are symmetrical point-to-point connections between the master unit and a designated slave unit. SCO connections use reserved time slots and can be considered as circuit-switched links between the master and slave units. They are mainly used to support time-sensitive information, such as voice calls. ACL connections send data packets in a directed manner and support both symmetrical and asymmetrical connections. In the reserved time slots of non-SCO connections, the master unit can establish packet-switched connections with any slave unit on a time slot basis. Bluetooth supports one asynchronous data communication channel, three synchronous voice channels, or one channel that simultaneously supports asynchronous data and synchronous voice. The voice channel has a rate of 64 kbit/s, and voice coding uses logarithmic PCM or Continuous Variable Slope Delta (CVSD) modulation. The rate of the asynchronous data communication channel is 723.2 kbit/s in the forward direction and 57.6 kbit/s in the reverse direction for asymmetric connections, and 433.9 kbit/s for symmetric connections.

3. Protocol Standards

The protocol structure of Bluetooth technology is shown in Figure 2.1.1. The entire protocol architecture is divided into three main parts: the underlying hardware module, the intermediate protocol layer, and the high-level application framework.

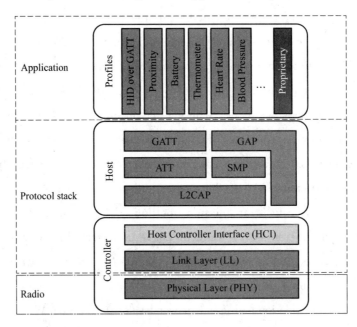

Figure 2.1.1　Protocol Structure of Bluetooth Technology

(1) Underlying Hardware Module

The underlying hardware module includes the Radio Frequency(RF), Baseband(BB), and Link Manager(LM) layers. The RF layer uses microwaves in the 2.4 GHz unlicensed ISM band to filter and transmit data streams. This layer protocol mainly defines the conditions that the Bluetooth transceiver needs to meet for normal operation in this frequency band. The BB layeris responsible for frequency hopping and the transmission of Bluetooth data and information frames. The LM layer is responsible for establishing and tearing down link connections while ensuring link security.

(2) Intermediate Protocol Layer

The intermediate protocol layer includes four items: L2CAP(Logical Link Control and

Adaptation Protocol), SDP (Service Discovery Protocol), RFCOMM (Radio Frequency Communication), and TCS(Telephony Control Protocol Spectocol). L2CAP mainly handles data fragmentation and protocol multiplexing, serves as the foundation for other upper-layer protocols. SDP provides a mechanism for upper-layer applications to discover available services and their characteristics in the network. RF-COMM, based on the ETSI standard TS07.10, simulates the functionality of a 9-pin RS232 serial port on top of L2CAP. TCS provides signaling for voice and data call control between Bluetooth devices. Between the BB and LM, there is also a Host Controller Interface (HCI) layer. HCI serves as the interface between software and hardware in the Bluetooth protocol, providing a unified command interface for accessing lower-level components, such as the BB, LM, status, and control registers.

(3) High-level Application Framework

The high-level application framework is located at the top of the Bluetooth protocol stack. Some typical application modes include dial-up networking, headset, LAN access, file transfer, etc.

Various applications can achieve wireless communication through their corresponding frameworks. The dial-up networking application mode can access the micro-network through RFCOMM simulated serial port. All devices connected through Bluetooth technology are considered as micro-network. A micro-network can be just two connected devices, such as a portable computer and a mobile phone, or eight devices connected together. In a micro-network, all devices are equal units with the same permissions. When the micro-network is initially established, one unit is defined as the master unit, and its clock and frequency-hopping sequence are used to synchronize the devices of other units. Other units are defined as slave units, and data devices can also access traditional LAN through them. Users can wirelessly transmit audio streams between their phones and earphones through the audio layer in the protocol stack. Multiple PCs or laptops can quickly and flexibly transfer files and share information without any wiring, and multiple devices can also achieve synchronized operations. With the continuous enhancement of mobile phone functions, mobile phone wireless remote control will also become one of the main application directions of Bluetooth technology.

4. Market Applications

Bluetooth technology connects various portable devices and cellular phones through wireless radio links, integrating computers with communication, and enabling people to exchange and transmit data and information anytime, anywhere. Bluetooth technology is not only applicable to home networks and small-scale offices, but also crucial for personal data communication. Therefore, the computer industry and the mobile communication industry attach great importance to Bluetooth technology, believing that it has a significant promoting effect on future wireless mobile data services. It is expected that wireless data communication services will grow rapidly in the coming years, and Bluetooth technology is considered one of the most significant advancements in wireless data communication. Among the five founding companies of Bluetooth, there are two well-known mobile communication manufacturers, two well-known portable computer manufacturers, and a leading company in chip technology and digital signal processing (DSP). After the announcement of their plan, it quickly gained support and adoption from many large companies, including Motorola, Lucent, Compaq, Siemens, Qualcomm, 3Com, TDK, and others. Since its establishment in May 1998, more than 1 300 companies have joined Bluetooth SIG. Each alliance member receives free patent licenses from other members to use Bluetooth technology in their products. Alliance members can access Bluetooth technology specifications and participate in discussions that are beneficial for product development and better compatibility among manufacturers.

Bluetooth technology has a wide range of applications. The Bluetooth headset introduced by Ericsson was the first Bluetooth product that people saw. Support for mobile phones and laptops was just the first stage of Bluetooth applications. It can be foreseen that in the next few years, mobile phone manufacturers will gradually launch mobile phones with Bluetooth capabilities. Then the application of Bluetooth will expand from handheld terminals to areas such as automobiles, aviation, consumer electronics, and information appliances. Currently, some manufacturers have developed several Bluetooth technology products for enterprises and ordinary consumers, including a portable hard drive called NetDrive. It uses Bluetooth technology to receive and store data wirelessly (with a total capacity of up to 200 MB). Computer users can perform wireless operations between the host and hard drive. When they leave, they can take a hard drive with them to prevent unauthorized access. When they return, they just need to reinstall the hard drive to continue working.

Car phones that support Bluetooth technology have also been developed. Car manufacturers are actively responding to Bluetooth technology and are planning to install hands-free phone systems in cars that work with Bluetooth-compatible mobile phones. Bluetooth can maintain wireless connections between mobile phones and personal computers. Even if the user's personal computer is in a suitcase, the user can receive emails through the phone and read the email titles on the mobile phone screen. Building a home network is one of the most important applications of Bluetooth technology. All information devices within the home are connected to form a home network, which is an inevitable trend in the future development of an information society.

2.1.2 Wi-Fi

1. The Evolution and Current State of Wi-Fi Technology

Wi-Fi technology is a technology that converts wired network signals into wireless signals. Almost all smartphones, tablets, and various smart home appliances support Wi-Fi Internet access, making it one of the most widely used wireless network transmission technologies today. Wi-Fi technology is usually used to establish a local area network and wirelessly connect various network devices. Through wireless transmission between these devices, network sharing, access, and control can be achieved.

Nowadays, Wi-Fi technology has become one of the main ways for billions of devices to access the Internet. The Wi-Fi technology-related standards were formulated by the Institute of Electrical and Electronics Engineers (IEEE) and are summarized in the 802.11 series.

In 1997, IEEE introduced the first 802.11 standard, which supported a theoretical maximum data rate of only 2 Mbit/s. However, the birth of this standard changed the way users access the Internet, freeing them from the constraints of cables.

To meet the demand for higher transmission speeds, IEEE released the 802.11b standard in 1999. 802.11b operates in the 2.4 GHz frequency band with a transmission rate of 11 Mbit/s, which is 5 times that of the original standard.

In the same year, IEEE also supplemented the 802.11a standard, which adopted the same core protocol as the original standard. It operates at a frequency of 5 GHz with a maximum original data transmission rate of 54Mbit/s, meeting the requirements for medium throughput (20 Mbit/s) in real-world networks. Due to the widespread use of the 2.4 GHz frequency band, using the 5 GHz frequency band gives 802.11a the advantage of fewer conflicts.

In 2003, the OFDM technology, which was originally part of the 802.11a standard, was adapted to operate in the 2.4 GHz frequency band, resulting in the creation of 802.11g. The carrier frequency of 802.11g is 2.4 GHz, with an original transmission speed of 54 Mbit/s and a net transmission speed of approximately 24.7 Mbit/s.

One of the most important standards that has had a significant impact on Wi-Fi is the 802.11n standard, which was released in 2009. This standard introduced major improvements to Wi-Fi transmission and access, including the introduction of MIMO, security encryption, and advanced features based on MIMO, such as beamforming and spatial reuse, enabling network transmission speeds of up to 600 Mbit/s. Additionally, 802.11n was the first Wi-Fi technology to operate simultaneously in both the 2.4 GHz and 5 GHz frequency bands.

With the rapid development of mobile services and the increasing demand for high-density access to Wi-Fi networks, the 802.11ac standard, released in 2013, introduced wider radio frequency bandwidth (up to 160 MHz) and higher-order modulation techniques (256-QAM), achieving transmission speeds of up to 1.73 Gbit/s and further improving the throughput of Wi-Fi networks.

However, as video conferencing, wireless interactive VR, ultra-high-definition video, smart homes, and mobile classrooms have become more prevalent, there is an increasing number of Wi-Fi access terminals. The development of IoT has also brought more mobile terminals accessing wireless networks. Therefore, Wi-Fi networks still need to continuously improve their speed while considering the ability to connect to more terminals and meet the increasing number of client devices and different application user experience requirements.

As early as 2014, the IEEE 802.11 working group began to address this challenge, and officially launched the 802.11ax standard in 2021. This standard introduces technologies such as uplink MU-MIMO, OFDMA frequency division multiplexing, and 1024-QAM high-order coding, providing a theoretical maximum speed of up to 9.6 Gbit/s. It also enhances energy efficiency, coverage, and security, meeting user access needs in device-intensive environments such as shopping malls, factories, and offices. In dense user environments, the average throughput of users has increased by at least 4 times compared to the previous generation, and the number of concurrent users has increased by more than 3 times. Therefore, 802.11ax is also known as High Efficiency Wireless (HEW).

Similar to previous releases of the new 802.11 standards, 802.11ax is also compatible with the previous 802.11ac/n/g/a/b standards, allowing old terminals to seamlessly connect to 802.11ax networks.

In order to enhance the recognition and awareness of Wi-Fi technology, in 2018, another important international organization in the Wi-Fi field, Wi-Fi Alliance (WFA), decided to adopt new naming rules to identify 802.11 standards, making it easier for users and device manufacturers to understand the devices they are connecting to or supporting. Taking inspiration from the naming conventions of cellular mobile communication 2G~5G, a numerical sequence was chosen to rename Wi-Fi. On the other hand, choosing a new generation of naming methods is also to better highlight the significant progress of Wi-Fi technology, which provides a wealth of new features, including increased throughput and faster speeds, support for more concurrent connections, etc. The corresponding generations of 802.11 standards are as Table 2.1.1.

Table 2.1.1 Historical 802.11/Wi-Fi Standards

Standard	Standard Approval Period	Maximum Bandwidth	Theoretical Maximum Speed
802.11ax/Wi-Fi 6	2021	160 MHz	9.6 Gbit/s
802.11ac/Wi-Fi 5	2013	160 MHz	6.9 Gbit/s
802.11n/Wi-Fi 4	2009	40 MHz	600 Mbit/s
802.11g	2003	20 MHz	54 Mbit/s
802.11a	1999	20 MHz	54 Mbit/s
802.11b	1999	20 MHz	11 Mbit/s
802.11	1997	20 MHz	2 Mbit/s

2. Technical Principles

In the 5G era, mobile data plans include more and more data at cheaper prices. However, it is still difficult to binge watch without any concerns. Home broadband, which is charged based on bandwidth, has unlimited data and can be converted into Wi-Fi signals through a wireless router. It not only provides Internet access for the whole family but also connects various smart home devices effortlessly.

Therefore, it is not an exaggeration to call the wireless router the data hub of the home.

The name "wireless router" can be divided into two keywords: wireless and routing.

Wireless refers to what is commonly called Wi-Fi. The wireless router can convert the home broadband from wired to wireless signals, allowing all devices to connect to their own Wi-Fi and enjoy Internet access. In addition, these devices form a wireless local area network (LAN) for high-speed data exchange, which is not limited by the bandwidth of the home broadband. For example, TV at home can be turned on directly through voice assistants. In fact, the voice assistant finds the TV through the LAN and sends commands without the need to connect to the Internet. However, if you want the TV to play the news, you will need to retrieve the data through the Internet.

As mentioned earlier, a local area network (LAN), also known as an internal network, is represented by LAN on a router. Therefore, Wi-Fi signals are also referred to as wireless LAN (WLAN). The Internet that we want to access is also known as the external network, represented by wide area network (WAN) on a router. Please refer to Figure 2.1.2 for details.

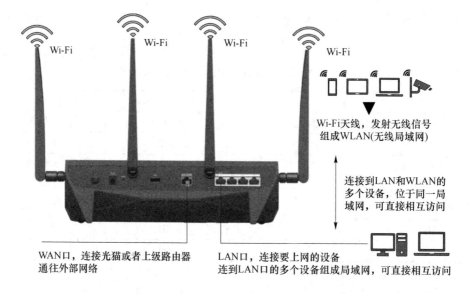

Figure 2.1.2 Wireless Router Interface Situation

In the internal network, each device has a different IP address, known as a private address. On the external network, all devices share the same public address, which is assigned by broadband service providers, such as China Telecom and China Unicom.

The router serves as a bridge between internal and external networks. The IP address translation and packet forwarding mentioned earlier are the routing functions of the router.

In general, the router is the hub of a home network, and all device data must pass through it for mutual access or to reach external networks. It is like a gatekeeper, and therefore a fully functional router is also known as a "home gateway".

Currently, Wi-Fi is the only mainstream technology for wireless local area networks (WLAN), so it can be considered equivalent to WLAN.

Wi-Fi primarily operates in two frequency bands: 2.4 GHz and 5 GHz. These two bands are known as the Industrial Scientific Medical (ISM) bands, and as long as the transmission power meets national standards, they can be used without authorization.

2.4 GHz is the earliest ISM band used globally, with a frequency range of 2.40 GHz to 2.483 5 GHz, and a total bandwidth of 83.5 MHz. Wi-Fi divides the 83.5 MHz bandwidth on the 2.4 GHz band into 13 channels, each with a 20 MHz width. These channels overlap, and originally only three could fit, but with the use of multicarrier modulation technology, 13 channels can be allocated. Interference between channels is difficult to avoid, so the network speed is slow, and there may be queuing for usage, as shown in Figure 2.1.3.

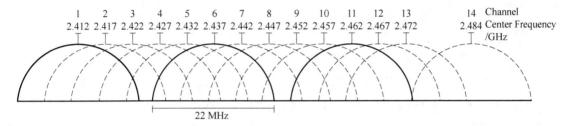

Figure 2.1.3　Wi-Fi Frequency Band Usage Chart

If the 2.4 GHz frequency band is like a narrow path, then the 5 GHz frequency band is undoubtedly a wide avenue.

The available range of the 5 GHz frequency band is from 4.910 GHz to 5.875 GHz, with a bandwidth of over 900 MHz, which is more than 10 times that of 2.4 GHz! This spectrum is too wide, so different countries have defined the range in which Wi-Fi can be used based on their own situations, as shown in Figure 2.1.4.

Figure 2.1.4

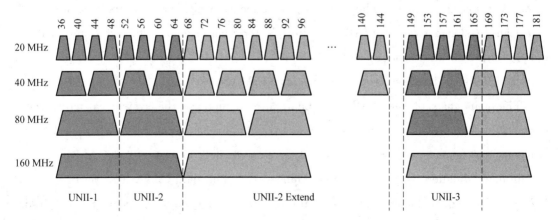

Figure 2.1.4　5 GHz Channel Distribution Chart

For example, in China, there are 13 20 MHz channels available for Wi-Fi in the 5 GHz spectrum, and continuous 20 MHz channels can also be combined to form 40 MHz, 80 MHz, or even 160 MHz channels.

The 5 GHz band has a large bandwidth and fewer devices running on it, so it naturally provides faster speeds and less interference. Therefore, if you want to achieve a good speed experience for your home network, you can consider using 5 GHz for full coverage. However, each coin has two sides. Although the 5 GHz band has a large bandwidth and less interference, the signal propagation attenuates quickly and it is easily blocked, resulting in weak wall penetration capabilities.

Wi-Fi 6 (802.11ax) is the latest standard for Wi-Fi technology. It is an improvement over the previous Wi-Fi 5 (802.11ac) standard and has several technological advantages, which are reflected in the following aspects:

(1) OFDMA Frequency Division Multiplexing Technology

Before Wi-Fi 6, the OFDM mode was used for data transmission, where users were differentiated by different time slots. In each time slot, a user occupies all subcarriers and sends a complete data packet. Wi-Fi 6 introduces a more efficient data transmission mode, OFDMA, which allocates subcarriers to different users and implements multiuser multiplexing of channel resources in the OFDM system.

(2) DL/UL MU-MIMO Technology

MU-MIMO uses the spatial multiplexing of channels to send independent data streams with the same bandwidth. Unlike OFDMA, where all users use the entire bandwidth, MU-MIMO provides multiplexing gain. Terminals are limited in terms of antenna quantity due to size, generally having only 1 or 2 spatial streams (antennas), which is fewer than the AP's spatial streams (antennas). Therefore, by introducing MU-MIMO technology in the AP, data transmission between the AP and multiple terminals can be achieved simultaneously, greatly improving the throughput.

(3) Higher-order Modulation Technology (1024-QAM)

The main goals of the Wi-Fi 6 standard are to increase the system capacity, reduce latency, and improve efficiency in high-density multiuser scenarios. However, better efficiency and faster speed are not mutually exclusive. 802.11ac uses 256-QAM orthogonal amplitude modulation, with each symbol transmitting 8 bit data ($2^8=256$). Wi-Fi 6 will use 1024-QAM orthogonal amplitude modulation, with each symbol transmitting 10 bit data ($2^{10}=1\,024$). The improvement from 8 to 10 is 25%, which means that compared to Wi-Fi

5, Wi-Fi 6's single spatial stream data throughput has increased by 25%.

(4) Spatial Reuse (SR) & BSS Coloring Mechanism

Wi-Fi 6 introduces a new mechanism for identifying same-frequency transmissions called BSS Coloring. It adds a BSS color field in the PHY header to "color" the data from different BSSs, assigning a color to each channel to identify a group of BSSs that should not interfere with each other. The receiving end can identify same-frequency interference signals early and stop receiving, avoiding wasting the transmission time. If the colors are the same, it is considered interference within the same BSS and the transmission will be delayed. If the colors are different, it is considered that there is no interference between them, and two Wi-Fi devices can transmit on the same channel and frequency in parallel.

(5) Expanding Coverage Range (ER)

Due to the Long OFDM symbol transmission mechanism adopted by the Wi-Fi 6 standard, the duration of each data transmission has been increased from the original 3.2 μs to 12.8 μs. The longer transmission time can reduce terminal packet loss. In addition, Wi-Fi 6 can use a minimum bandwidth of only 2 MHz for narrowband transmission, effectively reducing frequency-band noise interference, improving terminal sensitivity, and increasing coverage distance.

(6) Improving Battery Life

Target Wake Time (TWT) allows devices to negotiate when and how often they wake up to send or receive data. This feature can increase the device's sleep time and significantly extend the battery life of mobile and IoT devices.

(7) Security

The new WAP3 protocol has upgraded Wi-Fi security from four different dimensions. This includes upgrading the encryption level to 196 bit Advanced Encryption Standard (AES), rendering dictionary attacks ineffective, improving the security of IoT devices, and enabling personalized data encryption.

3. Protocol Standards

A wireless local area network (WLAN) consists of stations (STA), access points (AP), access controllers (AC), AAA servers, and network management units. The network reference model is shown in Figure 2.1.5. The AAA server is the entity that provides AAA services and supports the RADIUS protocol in the reference model. The portal server is used for authentication in web portals.

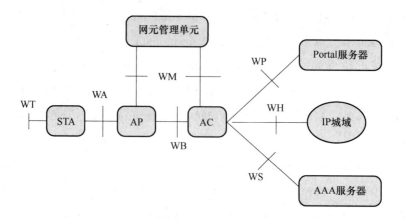

Figure 2.1.5　Wireless Local Area Network Reference Model

In this network model, the following interfaces are defined.

① WA Interface: The interface between the STA and the access point, also known as the air interface.

② WB Interface: The interface between the access point and the access controller. This interface is a logical interface and may not correspond to a specific physical interface.

③ WT Interface: The interface between the STA and the user terminal. This interface is a logical interface and may not correspond to a specific physical interface.

④ WU Interface: The interface between the Public Wireless Local Area Network (PWLAN) and the Internet.

⑤ WS Interface: The interface between the AC and the AAA server. This interface is a logical interface and may not correspond to a specific physical interface.

⑥ WP Interface: The interface between the AC and the Portal server. This interface is a logical interface and may not correspond to a specific physical interface.

⑦ WM Interface: The interface between public wireless LAN network element management units, which is a logical interface.

Network unit function:

① Station (STA) is the terminal in a wireless network. The STA accesses the AP through a wireless link, and the interface between the STA and the AP is the wireless interface WA.

② Access Point (AP) communicates with the STA through a wireless link, and the wireless link uses standard air interface protocols. Both the AP and the STA are addressable entities. In the uplink direction, the AP is connected to the AC through the wired interface WB.

③ Access Controller (AC) serves as a network management function between the wireless LAN and the external network. The AC aggregates data from different APs and connects to the Internet. The AC supports user security control, business control, billing information collection, and network monitoring. The AC can be directly connected to the AAA server or connected through an IP metropolitan area network backbone (supporting the Radius protocol). In specific network environments, the functions of the access controller AC and the access point AP can be physically integrated.

④ Authentication Authorization Accounting (AAA) servers provide authentication, authorization, and accounting functions. AAA servers can be physically composed of independent servers with different functions, namely Authentication Servers (AS), authorization servers, and accounting servers. The authentication server stores user authentication information and related attributes. When it receives an authentication request, it supports querying user data stored in the database. Once authentication is completed, the authorization server authorizes users with different attributes based on their information. In this standard, AAA servers support RADIUS protocol servers.

The portal server is responsible for pushing the PWLAN user portal website and is a mandatory network unit.

4. Market Applications

(1) Carrying High-bandwidth Videos Such as 4K/8K/VR

Video services are driving the development of ultra-wideband and gradually changing user behavior and demands. Operators are shifting their focus from connectivity to user experience. The bitrate of video services continues to increase, from standard definition to high definition, from 4K to 8K, and now to VR videos.

From the perspective of network transmission, the key factors affecting video quality include bandwidth, latency, and packet loss. Among them, bandwidth directly affects the video's bitrate, resolution, color depth, etc., which is the key elements of user experience. Packet loss directly affects the smoothness of the picture, which impacts the pleasure of user experience.

Wi-Fi 6 technology supports the coexistence of 2.4 GHz and 5 GHz frequency bands. The 5 GHz frequency band supports a 160 MHz bandwidth, with a maximum access rate of 9.6 Gbit/s. The 5 GHz frequency band has relatively less interference and is more suitable for transmitting video services. At the same time, it can reduce interference and packet loss through BSS coloring technology, MIMO technology, dynamic CCA, etc., to provide a better video service experience.

(2) Support for Low-latency Services Such as Online Gaming

Online gaming is considered a highly interactive service, where games need to respond promptly to user actions. In recent years, the rise of VR games and cloud VR games has placed higher demands on bandwidth and latency. For example, cloud VR games require a bandwidth of 80 Mbit/s~1 Gbit/s and a latency of less than 8~20 ms.

For VR games, the best current access method is through Wi-Fi wireless technology. Due to the large number of mobile devices and different types of services within a household, Wi-Fi is susceptible to various interferences. These factors combined can affect the gaming experience. Therefore, building a high-speed, low-latency, and reliable home wireless network is essential for enhancing the user's gaming experience.

(3) Smart Home and Intelligent Interconnection

Smart home and intelligent interconnection are important factors in smart home and smart security scenarios. Home interconnection technology mainly considers three aspects. First, whether it can connect enough devices. Currently, the number of sensor devices in smart homes can reach dozens or even hundreds. Second, whether the power consumption is low. Many smart devices are low-power devices that rely on batteries for long-term operation, such as smart locks. Third, whether the interoperability is user-friendly, and whether users can use commonly used terminals to control smart home devices.

The emergence of Wi-Fi 6 technology will bring about a unified technological opportunity for smart home connectivity. First of all, Wi-Fi 6 will become the mainstream of the next generation of Wi-Fi technology. Gateways, routers, phones, and some smart devices that support Wi-Fi 6 will appear like mushrooms after rain. At the same time, Wi-Fi 6 technology can adapt to dense and high-density access scenarios, enabling the Internet of Things within the home to be realized using Wi-Fi 6 and integrating the home IoT and home wireless LAN, making it easy for users to control home devices with any device, anywhere, and enhancing the user experience.

Wi-Fi 6 is suitable for home connectivity for another important reason. It borrows from the 802.11ah standard and introduces the Target Wake Time (TWT) function, which allows devices to negotiate when and how long they will be awakened before sending and receiving data. By allocating different TWT periods, the chance of competing for wireless media after awakening is reduced. TWT also sets a sleep time, which greatly extends the battery life for a large number of smart devices powered by batteries.

It can be seen that Wi-Fi 6 integrates high density, large quantity access, and low power consumption optimization, while also being compatible with various mobile terminals commonly used by users, providing good interoperability. Wi-Fi 6 is a promising technological choice in the field of smart home and intelligent interconnection in the future.

(4) Industry Applications

Using Wi-Fi 6 technology to build a wireless access network in the park can bring low cost, wide coverage, and high-quality mobile office access. Based on this, various video office collaboration services can be carried out to improve the employee network experience and enhance work efficiency.

Wireless access coverage in large indoor and outdoor public places, such as airport applications, is a typical high-density and intensive access public place. When providing Wi-Fi wireless access services to passengers, airports should not only consider network operation and maintenance management but also focus on the following aspects. The first aspect is how to achieve access for a large number of terminal users without reducing the overall efficiency of the wireless network. The Wi-Fi 6 standard solves the problems of network capacity and transmission efficiency from the aspects of spectrum resource utilization and multiuser access by introducing technologies such as uplink MU-MIMO, OFDMA frequency division multiple access, and 1024-QAM high-order coding. In a dense user environment, the average throughput of users is increased by at least 4 times compared to Wi-Fi 5, and the number of concurrent users is increased by more than 3 times.

The second aspect to consider is how to provide stable and high-quality wireless transmission to passengers. With the increasing number of video applications such as film and television, games, VR/AR applications, and mobile video offices, these services have higher performance requirements for network transmission: high bandwidth, low latency, and low error rates. Wi-Fi 6 improves the ability to resist interference and ensures stable and high-quality wireless access transmission, and enhances user experience through technologies such as narrowing the spacing between subcarriers, extending symbol length, BSS coloring, and dynamic CCA.

The third aspect to consider is how to provide secure access to passengers, especially in open environments. How to provide users with secure data access and transmission. Although the Wi-Fi 6 standard itself does not specify any new security features or enhancements, the Wi-Fi Alliance (WFA) has introduced a new generation of security encryption standards: WPA3, which is a more secure encryption method and has become a standard configuration for Wi-Fi 6. WPA3 enhances the security of user privacy through personalized data encryption for access to open networks, and encrypts the connection between each device and AP. Therefore, by adopting Wi-Fi 6 and WPA3 technologies, secure access can be provided to airport passengers.

2.1.3 ZigBee

1. 发展背景及现状

ZigBee 技术是基于 IEEE 802.15.4 标准的一种近距离、低复杂度、低功耗、低数据速率、低成本的无线通信技术。ZigBee 技术主要适用于自动控制和远程控制领域,可以嵌入各种设备,在工业控制、家庭智能化、无线传感器网络等领域有广泛的应用前景。ZigBee 这一名称来源于蜜蜂的 8 字舞,蜜蜂(Bee)是靠飞翔和嗡嗡(Zig)地抖动翅膀的舞蹈来与同伴传递花粉所在方位的信息,依靠这样的方式构成了群体中的通信网络。与之相应的 ZigBee 技术是一项开放性的技术,由各个国家、代工生产商、知名半导体生产商组成,组建了 ZigBee 联盟。该联盟是一个非营利组织,制定了基于 IEEE 802.15.4 协议,规范了可靠性强、性价比高、功耗低的网络应用规格。目前已有 150 多家成员公司正积极进行 ZigBee 规格的制定工作。ZigBee 联盟的主要目标是通过加入无线网络功能,为消费者提供更富弹性、更易用的电子产品。

Zigbee 技术主要有如下 8 个优点:

① 低功耗:在低耗电待机模式下,两节 5 号干电池就可支持 1 个节点工作 6~24 个月,甚至更长时间,这是 ZigBee 的突出优势。相较而言,蓝牙能工作数周,Wi-Fi 可工作数小时。

② 低成本:通过大幅简化协议(不到蓝牙的 1/10),降低了对通信控制器的要求。按预测分析,以 8051 的 8 位微控制器为例进行测算,全功能的主节点需要 32 KB 代码,子功能节点仅需要 4 KB 代码,而且 ZigBee 免协议专利费,每块芯片的价格大约为 2 美元。

③ 低速率:ZigBee 工作在 20~250 kbit/s 的较低速率,分别提供 250 kbit/s、40 kbit/s 和 20 kbit/s 的原始数据吞吐率,满足低速率传输数据的应用需求。

④ 近距离:传输范围一般介于 10~100 m 之间,在增加 RF 发射功率后,也可增加到 1~3 km。这是指相邻节点间的距离,如果通过路由和节点之间通信的接力,传输距离将可以更远。

⑤ 低时延:ZigBee 的响应速度较快,一般从睡眠转入工作状态只需 15 ms,节点连接进入网络只需 30 ms,进一步节省了电能。相较而言,蓝牙需要 3~10 s,Wi-Fi 需要 3 s。

⑥ 高容量:ZigBee 可采用星形、树形和网状网络结构,由一个主节点管理若干个子节点,

一个主节点最多可管理 254 个子节点。同时主节点还可由上一层网络节点管理,最多可组成 65 000 个节点的大网。

⑦ 高安全:ZigBee 提供了三级安全模式,包括无安全设定,使用访问控制列表(ACL)防止非法获取数据,以及采用高级加密标准(AES)的对称密码,以灵活确定其安全属性。

⑧ 免执照频段:采用直接序列扩频在工业科学医疗 2.4 GHz(ISM)频段,另外北美主要使用 915 MHz,欧洲主要使用 868 MHz。

2. 技术原理

简单地说,ZigBee 是一种高可靠的无线数据传输网络,类似于 CDMA 和 GSM 网络。ZigBee 数据传输模块类似于移动网络基站。通信距离从标准的 75 m 到几百米、几千米,并且支持无限扩展。ZigBee 是一个由多达 65 000 个无线数据传输模块组成的无线数据传输网络平台,在整个网络范围内,每一个 ZigBee 网络数据传输模块之间都可以相互通信。ZigBee 网络包含简单的点到点、点到多点通信,包装结构比较简单,主要由同步序言、数据、循环冗余校验(CRC)组成。ZigBee 网络采用数据帧的概念,每个无线帧包括了大量无线包装,包含了大量时间、地址、命令、同步等信息,真正的数据信息只占很小的一部分,而这正是 ZigBee 网络可以实现网络组织管理、实现高可靠传输的关键。同时,ZigBee 网络采用了介质访问控制技术(MAC)和直接序列扩频技术(DSSS),能够实现高可靠、大规模网络传输。

ZigBee 网络定义了两种物理设备类型:全功能设备(FFD)和精简功能设备(RFD)。一般来说,FFD 支持任何拓扑结构,可以充当网络协调器,能和任何设备通信。RFD 通常只用于星形网络拓扑结构中,不能完成网络协调器功能,且只能与 FFD 通信,两个 RFD 之间不能通信。RFD 的内部电路比 FFD 少,存储体容量较小,因此,其实现相对简单,也更利于节能。交换数据的网络包含 3 种典型的设备类型:协调器、路由器和终端设备。一个 ZigBee 网络由一个协调器节点,若干个路由器节点和一些终端设备节点构成。

协调器用于初始化 ZigBee 网络,它是网络中的第一个设备。协调器节点选择一个信道和一个网络标识符(又称 PAN ID),然后启动一个网络。协调器节点也可以用来在网络中设定安全措施和应用层绑定。协调器的角色主要是启动并设置一个网络,一旦该工作完成,协调器将以一个路由器节点的角色运行。由于 ZigBee 网络具有分布式的特点,故网络的后续运行不需要依赖于协调器的存在。

路由器的功能包括:允许其他设备加入网络,多跳路由和协助用电池供电的终端子设备的通信。路由器需要存储那些去往子设备的信息,直到其子节点醒来并请求数据。当子设备要发送信息时,子设备需要将数据发送给它的父路由节点。此时,路由器就要负责发送数据,执行任何相关的重发,如果有必要还需等待确认。这样,自由节点就可以继续回到睡眠状态。有必要认识到的是,路由器允许成为网络流量的发送方或者接收方。由于这种要求,路由器必须不断准备,以转发数据,因此它们通常需要用电线供电,而不是电池。如果某一工程不需要用电池来给设备供电,那么它可以将所有的终端设备作为路由器使用。

终端设备并没有维持网络的基础结构的特定责任,故它可以自己选择是休眠还是激活。终端设备仅在向它们的父节点发送数据或接收来自父节点的数据时才会激活。因此,终端设备用电池供电即可运行很长一段时间。与移动通信的 CDMA 或 GSM 网络不同的是,ZigBee 网络主要是为工业现场自动化控制数据传输而建立的,因此它必须具有简单、使用方便、工作可靠、价格低的特点;移动通信网络主要是为语音通信而建立,每个基站造价一般都在上百万人民币,而每个 ZigBee 基站造价却不到 1 000 元人民币。每个 ZigBee 网络节点不仅本身可以作为监控对象,例如,ZigBee 网络节点可以对其所连接的传感器直接进行数据采集和监控,还可以自动中转别的网络节点传过来的数据资料。除此之外,每一个 ZigBee 网络节点(FFD)还可以在自己信号覆盖的范围内和多个没有成为单网络的孤立子节点(RFD)建立无线连接。

3. 协议标准

相较于其他无线通信标准,ZigBee 技术协议架构显得更为紧凑和简单,大致分为底层硬件模块、中间协议层和高端应用层 3 部分,如图 2.1.6 所示。

用户应用程序		高端应用层
应用层(APL)(高级管理功能)		
设备配置(ZDC)子层	设备对象(ZDO)子层	
应用支持(APS)子层(提供ZigBee端点接口)		
网络层(建立和维护网络连接)		中间协议层
IEEE 802.15.4 LLC子层	IEEE 802.2 LLC子层	
	SSCS	
IEEE 802.15.4MAC子层(设备间数据传输)		
IEEE 802.15.4 868 MHz/915 MHz PHY	IEEE 802.15.4 2.4 GHz PHY	底层硬件模块
底层控制模块	ZigBee无线RF收发器	

图 2.1.6 ZigBee 技术协议架构

底层硬件模块是 ZigBee 技术的核心模块,所有嵌入 ZigBee 技术的设备都必须包含底层硬件模块。它主要由 ZigBee 无线 RF 收发器和底层控制模块组成。底层控制模块定义了物理无线信道和 MAC 子层之间的接口,提供物理层数据服务和物理层管理服务。物理层数据服务从无线物理信道收发数据,维护一个由物理层相关数据组成的数据库。数据服务主要包括如下 3 项。

① 信道能量检测:为网络层提供信道选择依据。它主要测量目标信道中接收信号的功率强度,检测结果为有效信号功率和噪声信号功率之和。

② 链路质量指示:为 MC 层或者应用层提供接收数据帧时无线信号的强度和质量信息。对信号进行解码,生成一个信噪比指标,并将信噪比指标和物理层数据单元一起提交给上层处理。

③ 空闲信道评估:判断信道是否空闲。ZigBee 技术协议标准定义了 3 种空闲信道评估模式。第一种是判断信道的信号能量,若信号能量低于某一个阈值,则认为信道空闲;第二种是判断无线信道的特征,这个特征主要包括两方面,即扩频信号和载波频率;第三种是前两种评估模式的综合,同时检测信号强度和信号特征,给出信道空闲判断。

中间协议层由 IEEE 802.15.4 MAC 子层、IEEE802.15.4 逻辑链路控制(LLC)子层、网络层,以及通过业务相关聚合子层(Service Specific Convergence Sublayer,SSCS)协议承载的 IEEE 802.2 LLC 子层(选用)组成。

① MAC 子层:使用物理层提供的服务实现设备之间的数据帧传输。LLC 子层在 MAC 子层的基础上,在设备间提供面向连接和非连接的服务。MAC 子层提供两种服务,MAC 层数据服务和 MAC 层管理服务。前者保证 MAC 协议数据单元在物理层数据服务中的正确收发,后者维护一个存储 MAC 子层协议状态相关的数据库。

② 网络层:负责建立和维护网络连接,独立处理传入数据请求、关联请求、解除关联请求和孤立通知请求。

③ SSCS 和 IEEE 802.2 LLC 子层只是 ZigBee 标准协议中可能的上层协议,并不在 IEEE 802.15.4 标准的定义范围内。SSCS 为 IEEE 802.15.4 的 MAC 子层接入 IEEE 802.2 标准中定义的 LLC 子层提供聚合服务。LLC 子层可使用 SSCS 的服务接口访问 IEEE 802.15.4,为应用层提供链路层服务。

高端应用层主要包括应用支持(APS)子层、设备对象(ZDO)子层、设备配置(ZDC)子层、应用层(APL)和用户应用程序。

① APS 子层:主要提供 ZigBee 端点接口。应用程序将使用该层打开或关闭一个或多个端点,并获取或发送数据。

② ZDO 子层:通过打开和处理目标端点接口来响应接收和处理远程设备的不同请求。与其他的端点接口不同,目标端点接口总是在启动时就被打开并被假设绑定到任何发往该端口的输入数据帧。

③ ZDC 子层:提供标准的 ZigBee 配置服务,定义和处理描述符请求。远程设备可通过 ZDO 子层请求任何标准的描述信息。当接收到这些请求时,ZDO 会调用配置对象,以获取相应的描述符值。

④ API 子层:提供高级协议管理功能。用户应用程序使用此模块来管理协议栈功能。

⑤ 用户应用程序:主要包括厂家预置的应用软件。同时,为了给用户提供更广泛的应用,该层还提供了面向仪器控制、信息电器和通信设备的嵌入式 API,从而可以更广泛地实现设备

与用户的应用软件间的交互。

4. 市场应用

ZigBee技术应用主要针对工业、家庭自动化、遥测遥控、汽车自动化、农业自动化和医疗护理等,例如,灯光自动化控制,传感器的无线数据采集和监控,油田、电力、矿山和物流管理等应用领域。除此之外,它还可以对局部区域内的移动目标进行定位,例如,城市中的车辆。具体应用领域如下:

① 监控照明、HVAC和写字楼安全;

② 配合传感器和激励器对制造、过程控制、农田耕作、环境及其他区域进行工业监控;

③ 带负载管理功能的自动抄表(AMR),这可使得地产管理公司削减成本和节省电气能源;

④ 对油气等生产、运输和勘测进行管理;

⑤ 家庭监控照明、安全和其他系统;

⑥ 对病患、设备及设施进行医疗和健康监控;

⑦ 军事应用,包括战场监视和军事机器人的控制;

⑧ 汽车应用,即配合传感器网络报告汽车的所有系统的状态;

⑨ 消费电子应用,包括对玩具、游戏机、电视、立体音响、DD播放机和家电设备进行遥控;

⑩ 用于计算机外设,例如,键盘、鼠标、游戏控制器及打印机;

⑪ 有源RFID应用,如电池供电标签,可用于产品运输、产品跟踪、存储较大物品和财务管理;

⑫ 基于互联网的设备之间的机器到机器(M2M)的通信。

2.2 低功耗远距通信

2.2.1 Sigfox

1. 发展背景及现状

随着5G的飞速发展,其"对立面"——低功耗广域网络(LPWAN)也逐渐被广泛应用。5G具有更大的带宽、更快的数据传输速率以及更大的数据吞吐量,但与之对应的是更高的成本,更多对频谱资源的占用以及对空中无线资源的占用。而在很多时候,我们不需要如此高的数据传输量,也不需要付出如此高的成本。以Sigfox为例,其一套成熟的通信模块的成本大约为 3～5 美元，在 5G 应用中，仅仅一个基带芯片的成本就远高于此。

智慧城市、智慧道路、智慧医疗、智慧物流等很多领域，往往不需要音频、视频等大数据量的传输，其传输的往往是终端传感器中采集的少量数据以及一些简单的数据分析。由此可见，LPWAN 和 5G 这两种无线通信类型将在物联网这个大的世界中扮演各自的角色，并最终都将汇合在万物互联的尽头。

Sigfox 技术是低功耗广域网（LPWAN）的主流技术之一。

Sigfox 技术是由法国的一家名叫 Sigfox 公司推广的，Sigfox 公司成立于 2009 年，总部位于法国图卢兹。Sigfox 技术通过组建一个低功耗的广域网来为工业、物流业和能源业等行业应用提供低成本的物联网链路。

该技术运行频率为 902 MHz，可以为智能电表和其他不需要太多带宽且数据传输量较小的数十亿设备提供网络连接。Sigfox 技术的优势在于其成本要比采用授权频谱的蜂窝物联网技术低得多。Sigfox 芯片在部分市场可以以每年每台设备低至 1 美元的价格提供连接服务。

Sigfox 一度有希望成为物联网市场的核心技术提供者。早期，Sigfox 技术覆盖了美国 20% 的人口，100 多个美国城市。2017 年时，Sigfox 称已在超过 36 个国家提供网络连接，包括法国 Orange、西班牙电信（Telefonica）和欧洲有线电视公司 Altice 等运营商都宣布与 Sigfox 达成了合作交易。然而，Sigfox 技术仅在全球范围内支持 2 000 万个连接，这比之前的预期目标要小得多。

在 LPWAN 的技术实践中，Sigfox 是面世最早的技术，但近几年 Sigfox 的市场占有率却在逐步降低，NB-IoT 和 LoRa 逐渐成为市场主导，LoRa 更是成为全球部署数量最多的解决方案。

Sigfox 与 LoRa 差距日渐拉大的主要原因在于其不同的商业模式。与全球范围内的广域物联网连接数相比，Sigfox 的份额微乎其微。但 Sigfox 技术拥有后端数据、云服务及终端软件的全部技术，其作为一种低功耗广域网络技术仍然具有一定的潜力。

2022 年 4 月，新加坡物联网公司 UnaBiz 收购了破产的 Sigfox。UnaBiz 表示将推动 Sigfox 成为一个更加开放的标准。UnaBiz 已重新确立了 Sigfox 经营策略及其商业模式，使其更聚焦于为重点行业提供服务，在智能计量、设施管理、资产追踪、资产管理等方面推动可大规模部署的定制化物联网解决方案落地。截至 2022 年底，Sigfox 的亏损额比 2021 年底大幅减少了 2/3。

目前，Sigfox 正融合其他 LPWA 技术和小无线通信技术进行服务（此前已促成 Sigfox 与 LoRaWAN® 的技术整合），以推动物联网技术互操作性和 LPWAN 统一。Sigfox 网络和技术在全球多个区域重新活跃起来。

UnaBiz 已与腾讯、华普微电子、纵行科技、深科技等国内企业进行深入交流，在芯片、模组、连接互操作等方面探讨合作意向。未来，中国市场和合作伙伴或将成为 UnaBiz 推动

Sigfox生态的重要载体。

2. 技术原理

LTN(低吞吐量网络)的网络架构如图2.2.1所示。

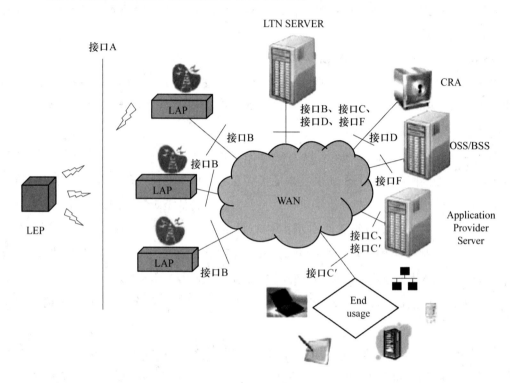

图2.2.1 LTN网络架构

LTN网络架构各接口说明如下。

接口A：在LEP(终端设备)和LAP(基站或网关)之间使用无线接入技术。LEP代表LTN端点，LAP代表LTN接入点。

接口B：通过WAN介质在LAP和LTN服务器之间使用。广域网介质可以是ADSL、光纤、卫星或微波链路。

接口C：在LTN服务器和应用程序提供商服务器之间使用，使用IP协议。

接口D：在LTN CRA(中央注册机构)和LTN服务器之间使用。

接口E：在多个LTN服务器之间使用，在漫游时使用。

接口F：用于LTN服务器和OSS/BSS服务器之间，主要功能是交换注册和网络状态的数据。

接口A′：在LEP内部使用。它用于DCS(Data Collection System)和LTN模块之间，通过串行连接的AT命令实现。

接口C′：用作最终用户界面，由应用程序提供商提供。

接口 F′：用于应用程序提供程序和 OSS/BSS 服务器之间。

Sigfox 的网络架构是从 LTN 的架构演变而来。Sigfox 网络由对象（最终用户设备）、Sigfox 网关或基站、Sigfox 云和应用服务器组成。

Sigfox 的网络架构如图 2.2.2 所示。

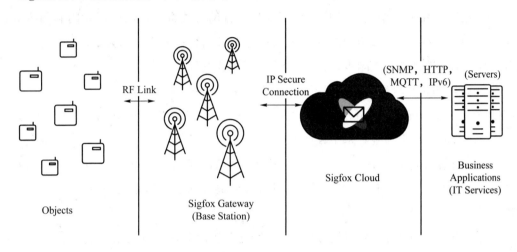

图 2.2.2　Sigfox 网络架构

Sigfox 对象使用星形拓扑与网关连接，Sigfox 网关和 Sigfox 云之间由直接的安全的点对点传输。云与服务器连接时可以使用不同的协议，如 SNMP、MQTT、HTTP、IPv6 等。

Sigfox 的上行帧结构如图 2.2.3 所示，上行链路 MAC 帧中包含：前同步码（4 B）、帧同步（2 B）、端点 ID（4 B）、有效载荷（0～12 B）、认证（可变）、FCS（2 B，用作 CRC）。

preamble	frame sync.	end-device ID	payload	auth.	FCS
4 B	2 B	4 B	0~12 B	可变	2 B

图 2.2.3　Sigfox 上行帧结构

Sigfox 的下行帧结构如图 2.2.4 所示，下行链路 MAC 帧中包含：前同步码（32 bit 或 4 B）、帧同步（13 bit）、标志（2 bit）、FCS（8 bit）、认证（16 bit）、错误代码（可变）、有效负载（可变）。

preamble	frame sync.	flags	FCS	auth.	error codes	payload
32 bit或4 B	13 bit	2 bit	8 bit	16 bit	可变	可变

图 2.2.4　Sigfox 下行帧结构

Sigfox 对基带的处理为超窄带（Ultra-Narrow Band，UNB）技术，使用 192 kHz 频谱带宽

的公共频段来传输信号,每条信息的传输宽度为 100 Hz,以 100 bit/s 或 600 bit/s 的数据速率传输,具体速率取决于不同区域的网络配置。同时,UNB 技术使 Sigfox 基站能够远距离通信,不容易受到噪声的影响和干扰。

Sigfox 系统通信流程的特点包括随机接入、协作接收、短消息通信、双向传输。

随机接入:网络和设备的消息传输采用异步传输的方式进行,设备以"时间和频率分散"(Time and Frequency Diversity)的方式发送消息,即设备以随机选择的频率发送消息,然后再以不同的频率发送另外两个副本,以保证消息发送的完整性。

协作接收:与传统蜂窝网络不同,任何终端设备都不附着在特定的基站上,Sigfox 系统采用"空间分散"(Spatial Diversity)的方式接收上行消息,即设备发送消息可以由任何附近的基站接收,实际部署中,平均接收基站数量为 3 个。

短消息通信:Sigfox 设计了一个特殊的短消息通信协议,消息的大小为 0~12 B,足以传输传感器数据、状态、警报、GPS 坐标等事件,以解决低成本的远距离覆盖和终端设备低功耗的问题。

双向传输:下行消息由终端设备触发。设备发送触发下行消息的第一帧 20 s 后,将有一个最长持续时间为 25 s 的接收窗口,Sigfox 云服务器接收到设备发送的带有下行触发标识的消息后,会协商客户服务器发送下行消息。

3. 协议标准

Sigfox 的协议栈由射频层、物理层、MAC 层和应用层组成,如图 2.2.5 所示。

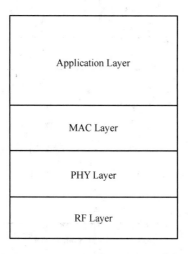

图 2.2.5 Sigfox 协议栈结构

其中,射频层有两种实现方式,分别为超窄带(UNB)和正交序列扩频(Orthogonal Sequence Spread Spectrum,OSSS),在世界不同国家和地区分配了不同的频谱,如美国使用 915 MHz 频段、欧洲使用 868 MHz 频段、中国使用 433 MHz 频段。

物理层主要负责在传输和接收期间处理MAC帧,系统在上行链路中使用BPSK调制,在下行链路中使用GFSK(在UNB实现中)调制。终端与网络的通信使用插入前导码的方式进行同步,即终端发送给网关的上行消息会添加前导码,通知对方做好接收准备,网关发送给终端的下行消息会将前导码删除。

MAC层会根据定义的上行链路和下行链路结构处理MAC帧的封装和解封装,上行链路消息通过UNB传输,下行链路消息的传输可以使用带内或广播来实现。除此之外,MAC层还需要完成终端用户的身份验证和使用FCS(帧检验序列)进行错误检测。

应用层支持多种接口协议,如SNMP、HTTP、MQTT、IPv6等,可以根据用户需求定义不同的应用程序,如Web、消息等。

4. 市场应用

Sigfox在物联网中的应用场景比较广泛,常见的应用包括以下10种。

① 智能停车场:通过在车位上安装传感器,Sigfox可以实时监测车位的使用情况,从而提供实时的停车位信息,方便司机快速找到空闲的车位。

② 智能农业:通过在农田中安装传感器,Sigfox可以实时监测土壤湿度、温度、光照等信息,例如,土壤湿度传感器可以实时监测土壤湿度,帮助农民合理浇水,从而帮助农民更好地管理农田,提高农作物的产量和质量。

③ 智能物流:通过在货物上安装传感器,Sigfox可以实时监测货物的位置、温度、湿度等信息,例如,货物跟踪传感器可以实时监测货物的位置和状态,提供实时的物流信息,从而提高物流的效率和安全性。

④ 智能城市:通过在城市中安装传感器,Sigfox可以实时监测交通、环境、能源等信息,监测和管理城市设施,例如,垃圾桶传感器可以实时监测垃圾的填充情况,优化垃圾收集路线,从而提高城市的管理效率和居民的生活质量。

⑤ 智能健康:通过在身体上安装传感器,Sigfox可以实时监测身体的健康状况,例如,心率传感器可以实时监测人体的心率,并将数据发送到云端进行分析,以帮助人们及时发现健康问题,从而提高健康管理的效率和精度。

⑥ 智能家居应用:通过在室内安装传感器,Sigfox可以实时监测空气质量和室内环境,例如,温度和湿度传感器可以实时监测室内环境,智能控制系统可以自动调节温度和湿度,提供舒适的居住环境。

⑦ 资产追踪:通过在物体上安装传感器,Sigfox可实时监测物体的状态和位置,例如,人们可以将Sigfox设备安装在贵重物品上,实时监测物品的位置,防止丢失或盗窃。

⑧ 环境监测:通过在化工厂安装传感器,Sigfox可以实时监测工厂的生产环境和排放物情况,例如,空气质量传感器可以实时监测空气中的污染物浓度,提供实时的环境质量信息。

⑨ 安防监控：通过在居民楼安装传感器，Sigfox 可以实时监测楼内外入户门的状态，例如，门窗传感器可以实时监测门窗的状态，一旦发生异常，系统会发送警报信息，提醒用户注意安全。

⑩ 物联网支付：通过在移动终端安装传感器，Sigfox 可以实时提供支付、交易功能，例如，智能支付终端。

2.2.2　NB-IoT Technology

1. Development Background and Current Situation

At the beginning of the NB-IoT technology standardization work, there were significant differences in the uplink transmission schemes in the industry, resulting in slow standardization progress. In order to ensure that the NB-IoT standard could be completed as scheduled, China Mobile took the lead in organizing multiple vendors' uplink transmission schemes into a unified technical solution, greatly promoting the smooth completion of standardization. In addition, China Mobile, together with other partners, conducted in-depth research on technical solutions related to new features of the Internet of Things, such as narrowband IoT wide coverage, massive connections, low power consumption, and low cost. They completed technical solutions with NB-IoT features, including flexible deployment scenarios, ultra-narrowband system design, control plane data transmission, deep sleep, etc., and pushed them into the 3GPP standard. For issues discovered during field testing, such as delayed reselection, lack of support for connection reestablishment, lack of support for measurement reporting, and failure to consider downlink channel quality in scheduling, solutions were proposed and written into the 3GPP standard.

On June 16, 2016, the NB-IoT technology protocol was approved by the 3GPP Radio AccessNetwork (RAN) Technical Specification Group meeting. The NB-IoT specification was frozen, and the standardization work was completed. The NB-IoT standard went from project initiation to protocol freezing in less than 8 months, making it one of the fastest established 3GPP standards in history. In the first quarter of 2017, according to the "National Next Generation Information Technology Industry Plan," the NB-IoT network was designated as one of the key projects in the information and communication industry's "Thirteenth Five-Year Plan." In June 2017, the General Office of the Ministry of Industry and Information Technology officially issued the "Notice on Comprehensive Promotion of Mobile Internet of Things (NB-IoT) Construction and Development," clarifying the significance of building and developing NB-IoT networks. In December 2017, the Global Mobile Communications Equipment Suppliers Association released data showing that there were already 31 commercial networks worldwide. By June 2018, there were already 46 commercial networks worldwide. By December 2018, there were already 89 commercial networks worldwide. These data all indicate that the development of the Internet of Things based on NB-IoT technology is accelerating globally, and the development process of NB-IoT can be roughly summarized as the incubation stage, the standardization stage, and the beginning of application.

In terms of network deployment, the three major domestic telecommunications operators in 2017 actively laid out the mobile Internet of Things. Network construction entered an accelerated phase in the third quarter, and by the end of 2017, the number of NB-IoT base stations nationwide exceeded 400 000. It is estimated that by 2020, the scale of NB-IoT base stations in China will reach 1.5 million, and the large-scale deployment of NB-IoT networks will bring ubiquitous coverage for the mobile Internet of Things. As network construction gradually completes, the three major operators have also successively announced the commercial use of NB-IoT nationwide in 2017.

In terms of business applications, rapid development is seen in areas such as municipal services, smart buildings, transportation and logistics, including water, electricity, and gas meter reading, smart parking, public housing renovation, smart fire protection, smart trash cans, environmental monitoring, smart manhole covers, smart street lights, smart landscapes, and shared bicycles.

2. Technical Principles

The NB-IoT network system architecture, as shown in Figure 2.2.6, consists of wireless network, core network, business platform (such as IoT platform), application server (AS), basic communication suite, and terminals and user cards.

(1) Core Network

The core network consists of packet domain core network, circuit domain core network, and user database (HSS). The packet domain core network consists of MME, S-GW, P-GW, and other functional units, which mainly provide packet domain data transmission and capability opening functions. The circuit domain core network consists of MSC/VLR, short message center (SMSC), and other functional units, which mainly provide short message transmission functions. The user database includes HSS, etc.

(2) Function Business Unit Platform, Mainly Providing User Contract Data Functions

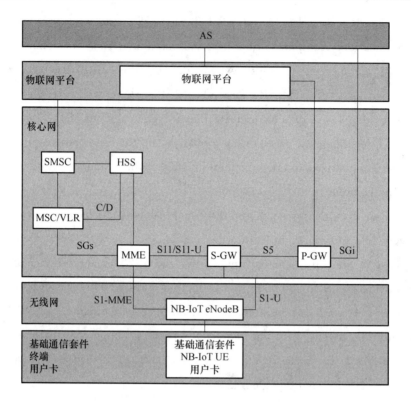

Figure 2.2.6 Basic System Structure of NB-IoT Network

The business platform includes multiple platforms, such as the Internet of Things capability open enablement platform, connection management platform, over-the-air card writing platform, and business gateway for users. In order to open network capabilities to users, the Internet of Things capability open enablement platform provides terminal devices with device access, data storage, data routing, and forwarding functions, and offers data push, device management, data query, command issuance, and other functions to upper-layer applications.

The connection management platform is a customer-oriented operation support platform that provides services such as user card information queries, communication management, and data statistics and analysis to customers.

The over-the-air card writing platform can dynamically switch between user numbers, that is, It can change numbers without changing physical cards. The platform has functions such as single card key generation for card production, batch card writing, card data management, number management, and card writing log management.

The business gateway is a business layer access device that provides communication access, business authentication, message routing, protocol conversion, and other functions for terminals, business platforms, and capability systems in the Internet of Things business system.

(3) Wireless Network

NB-IoT wireless access network (E-UTRAN) consists of eNodeB, which allows users to access the network through the air interface and connect the signaling interface to the MME through the S1-MME interface and the user interface to the S-GW through the S1-U interface.

(4) Terminals and User Cards

NB-IoT terminals need to meet technical requirements such as basic functions, communication functions, business functions, RF performance, card interface capabilities, and electromagnetic compatibility, and they must have NB-IoT network access capabilities.

As a user identification, the user card mainly provides functions such as storing secure data, dynamically loading and updating applications, secure computation, identifying user identity, and authentication access.

(5) Basic Communication Suite

A communication middleware for NB-IoT terminals, which realizes functions such as communication between IoT terminals and business platforms, application data transmission, and device management. It is an important component to enhance data service capabilities.

The application server completes functions such as storing, forwarding, and managing vertical industry-related data.

NB-IoT is a new technology designed specifically for LPWA mobile IoT services. It adopts the concept of ultra-narrowband and low-cost design (based on FDD half-duplex mode), sacrificing rate and latency performance to achieve more extreme IoT characteristics such as low power consumption and large-scale connections. The design goals of NB-IoT include: 10-year battery life, 20 dB coverage enhancement compared to GSM systems, support for 50 000 connections per small area, and lower cost. The new air interface technologies used include: ultra-narrowband design (200 kHz system bandwidth, 3.75 kHz/15 kHz channel bandwidth), coverage enhancement through repeated transmission; added power-saving modes to reduce power consumption (such as PSM power-saving mode, eDRX function, etc.); low complexity design to reduce costs.

The subcarrier spacing of LTE is 15 kHz. In order to further adapt to small packet transmission for typical IoT services, NB-IoT introduces a smaller subcarrier spacing in the uplink, such as 3.75 kHz. The uplink data transmission scheme includes 3.75 kHz single frequency, 15 kHz single frequency, and 15 kHz multi-frequency schemes. For 3.75 kHz single frequency, the minimum transmission unit in the time domain is 32 ms; for 15 kHz single frequency, the minimum transmission unit in the time domain is 8 ms; for 15 kHz multi-frequency, NB-IoT system supports 3-frequency, 6-frequency, and 12-frequency. Among them, for 15 kHz 3-frequency, the minimum transmission unit in the time domain is 4 ms; for 15 kHz 6-frequency, the minimum transmission unit in the time domain is 2 ms; for 15 kHz 12-frequency, the minimum transmission unit is the same as LTE, which is 1 ms in the time domain. The single user uplink peak rate of 3.75 kHz single frequency is about 4.7 kbit/s; the single user uplink peak rate of 15 kHz single frequency is about 15.6 kbit/s; the single user uplink peak rate of 15 kHz multi-frequency is about 62.5 kbit/s. The downlink subcarrier spacing remains the same as LTE, with only one subcarrier spacing (15 kHz). The minimum transmission unit in the time domain is 1 ms, which is the same as LTE. In this design, the downlink peak rate is about 21 kbit/s.

NB-IoT systems can support three working modes to adapt to different scenarios of different operators. The three working modes include Stand-alone mode, which uses independent spectrum deployment for NB-IoT; In-band mode, which deploys NB-IoT within the LTE bandwidth and is suitable for operators who only have LTE spectrum without additional spectrum; and Guard band mode, which deploys NB-IoT within the guard band at the edge of the LTE spectrum and is suitable for operators who only have LTE spectrum without additional spectrum.

3. Protocol Standards

The overall protocol architecture of NB-IoT is shown in the following Figure 2.2.7.

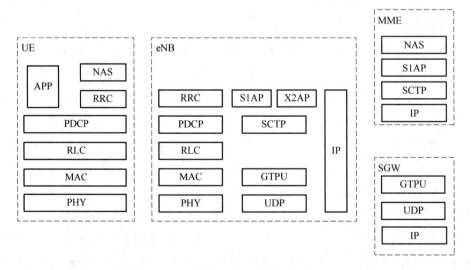

Figure 2.2.7 NB-IoT Overall Protocol Structure

Different from the traditional LTE/EPC architecture, NB-IoT supports control plane optimization schemes that involve significant modifications and enhancements to the protocol stack. The control plane optimization schemes include two types:

① Control plane optimization scheme based on SGi, as shown in the following Figure 2.2.8.

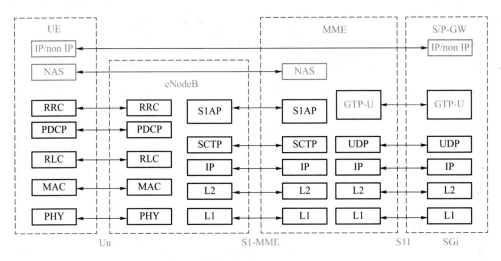

Figure 2.2.8　NB-IoT Control Plane Optimization Scheme Based on SGi

② Control plane optimization scheme based on T6, as shown in the following Figure 2.2.9.

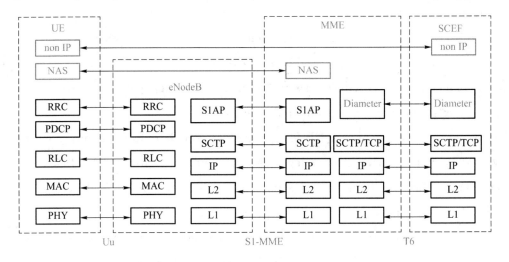

Figure 2.2.9　NB-IoT Control Plane Optimization Scheme Based on T6

From the above protocol stack, it can be seen that:

a. The IP data/non-IP data packets of the UE are encapsulated in NAS packets.

b. The MME performs the conversion of NAS packets to GTP-U packets. For uplink small data transmission, the MME extracts and re-encapsulates the IP data/non-IP data packets from the NAS packets into GTP-U packets and sends them to the SGW. For downlink small data transmission, the MME extracts the IP data/non-IP data from the GTP-U packets, encapsulates them in NAS packets, and sends them to the UE.

The user plane optimization solution does not modify or enhance the LTE/EPC protocol stack. For the user plane optimization solution, on the Uu interface, the NB-IoT terminal can initiate the RRC connection suspension (RRC Suspend) and RRC connection resumption (RRC Resume) processes. Unlike entering idle mode, after the RRC connection is suspended, the key information of the UE's access layer context is still saved on the UE and eNodeB. When the NB-IoT terminal initiates the RRC connection resumption, the eNodeB can quickly rebuild the RRC connection, restore the wireless air interface bearer previously allocated to the UE, and restore the S1 connection, thereby quickly restoring the uplink data transmission channel.

4. Market Applications

NB-IoT is mainly aimed at IoT services that require wide coverage, massive connections, low power consumption, low data rates, and ultra-low costs. Examples include abnormal reporting services (smoke alarm detectors, smart meter power outage notifications, gas meter gas outage notifications, etc.); periodic reporting services (smart meter reading, logistics tracking, smart homes, smart city IoT, smart buildings, etc., reporting data in hourly, daily, weekly cycles, with a small data volume each time); network command services (device switch notifications, trigger data reporting, software upgrades, etc.); and software upgrade push, etc.

With the completion of the NB-IoT technology standard formulation, domestic and foreign operators have conducted in-depth research and exploration in NB-IoT technology testing, verification, and industrial construction, and have taken the lead in applying it in several major areas such as smart city management, smart logistics, industrial IoT, smart wearables, smart homes, and wide-area IoT. The combination of NB-IoT technology with traditional industries and products will help improve the battery life of consumer electronics products (such as wearable devices), assist in the IoT upgrade and transformation of infrastructure and traditional industries, realize intelligent management of cities, factories, and homes, and precise maintenance of devices, and provide more effective solutions for key areas such as environmental monitoring and landslides, achieving social benefits of cost reduction and efficiency improvement.

Currently, smart water meters have been commercially used in many places, such as Yingtan, Jiangxi, playing a demonstrative role in the application of NB-IoT in the meter reading market.

Based on NB-IoT, smart street lights have been widely piloted and applied in many cities in China, including Beijing, Nanjing, Hangzhou, and Yingtan. Taking Beijing Future Science City as an example, the application of NB-IoT smart street lights will be introduced.

Beijing Future Science City is a key implementation project of the national "Thousand Talents Program", covering an area of 16 km^2. It has attracted many central enterprises such as China National Offshore Oil Corporation, State Grid Corporation of China, China Electronics Corporation, Aluminum Corporation of China, and COFCO Group. It is a comprehensive new type of park that integrates high-end talents, scientific and technological resources, and brand effects. It attaches great importance to the construction of various aspects of the park and takes the lead in introducing the latest technology to create a world-class talent innovation and entrepreneurship base. Among them, the lighting project in the waterfront park is completed by a company in Zhejiang, using the NB-IoT smart street light solution.

Currently, shared bicycles based on NB-IoT have been deployed in all cities nationwide for many years and have received a good user experience.

2.2.3 LoRaWAN Technology

1. Development Background and Current Situation

We are familiar with LoRa technology, which is a wireless communication technology developed by Semtech. It uses spread spectrum modulation to expand the signal in the frequency domain, allowing the signal to be received at a lower signal-to-noise ratio. LoRa technology is a physical layer technology known for its low power consumption, long distance, and strong anti-interference capability.

LoRaWAN technology, on the other hand, is a set of protocol standards based on the LoRa physical layer transmission technology, with the MAC layer as the main component. It plays the role of a network management protocol for LPWAN (Low Power Wide Area Network) and is responsible for sending information between gateways and end nodes. Its main features are low power consumption, long distance, and wide area coverage.

LoRaWAN is the most popular and widely deployed protocol for LoRa wide area networks.

LoRaWAN was developed in 2009 by Grenoble-based company Cycléo, which was acquired by Semtech in 2012. Three years later, Semtech established the LoRa Alliance, which currently has over 500 members including IoT product and service providers, manufacturers, and telecommunications companies. Semtech promotes the protocol and develops its applications worldwide through this LoRa Alliance. In 2021, LoRaWAN became one of the officially recognized Low-Power Wide-Area Network (LPWAN) standards by the International Telecommunication Union (ITU) as a Medium Access Control (MAC) layer protocol. LoRaWAN uses unlicensed radio spectrum in the ISM (Industrial, Scientific, and Medical) bands, with frequencies around 900 MHz or 430 MHz (exact frequencies vary worldwide). Using unlicensed spectrum means that companies can easily deploy networks and provide dedicated networks for enterprises.

The main characteristics of LoRaWAN are as follows.

① Low power consumption and long service life: LoRaWAN data transmission and reception require low current (less than 50 mA), greatly reducing the power consumption of devices. A single charge can provide a service life of up to ten years, significantly reducing support and maintenance costs.

② Cost savings: The wide coverage range and relatively low gateway cost significantly reduce the deployment cost of LoRaWAN networks. For devices, the price of communication modules is within the range of $10, and the cost of connectivity is only $1 per year due to unlicensed spectrum.

③ Location Services: Since signals from specific devices can be received by multiple gateways, the device's location can be calculated based on the signal strength and/or signal arrival time of each base station, enabling network-based location services that can be used for tracking or geofencing of devices.

④ Deep Penetration: LoRa radio modulation allows for deep indoor penetration and increases the ability to reach sensors located underground, such as water or gas meters.

⑤ No need to obtain any frequency licenses.

⑥ Fast Deployment and Commercialization: The LoRaWAN open standard, combined with cost-free operating frequencies and low-cost base stations, allows operators to launch networks with minimal investment in just a few months. Two-way communication fully supports various use cases that require both uplink and downlink, such as street lighting, smart irrigation, energy optimization, or home automation.

⑦ One-stop management: LoRaWAN network supports various vertical solutions, allowing service providers to use one platform and standard to manage various use cases, such as smart buildings, precision agriculture, smart metering, or smart cities.

2. Technical Principles

As shown in Figure 2.2.10, the LoRaWAN network system consists of four parts: end nodes, gateways, network servers, and application servers. Among them, end nodes are also called terminal devices, sensors, or nodes; gateways can also be called concentrators or base stations, corresponding to base stations in mobile communication networks; network servers correspond to mobile switching centers (MSC) or mobile-assisted handover (MAHO) in mobile communication networks. Depending on the application and service, the LoRaWAN network system requires the support of application servers, which are necessary components of the LoRaWAN system, but not necessary components in mobile communication networks. LoRaWAN nodes are always present to meet certain business needs, and there are almost no cases where they are connected to the network without any service business, while the purpose of mobile communication services is to keep users connected to the network and does not care about what business they are running. This is also one of the important differences between the network system of the Internet of Things and the mobile communication network system.

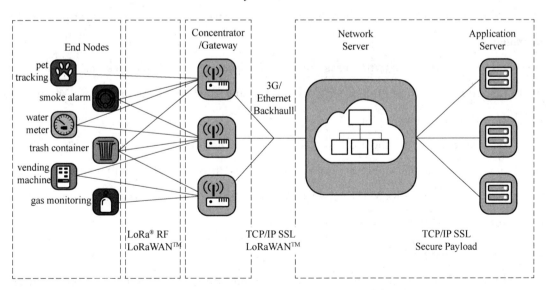

Figure 2.2.10　The System Composition of LoRaWAN Network

In the LoRaWAN network, the end nodes are connected to the gateway through LoRa wireless communication. The gateway is connected to the network server through existing wired/wireless networks (Ethernet/cellular network), and the network server is further connected to the application server through Ethernet. The communication process can be initiated by the end node or initiated by the application server. The gateway and network server are responsible for transparent transmission and network management, without direct involvement in the business. The LoRaWAN gateway continuously receives data from the nodes and forwards it to the network server, which then organizes the data and sends it to the application server. Upon receiving the business data from the nodes, the application server responds with instructions and sends them to the network server, which manages the gateway to deliver the commands to the original business nodes.

From the system of the LoRaWAN network, we can see the characteristics of network connection.

① The terminal node is connected to a LoRaWAN gateway. Some terminal nodes (such as the water meter in the diagram) are connected to two gateways, while others (such as the trash can in the diagram) are connected to three gateways. This is completely different from the way a user connects to a gateway in a mobile communication network system.

② All gateways in the LoRaWAN network system are independent and have no relationship with each other. There is no need for frequency allocation or cross-network management.

③ The LoRaWAN network uses traditional wired or 3G/4G networks as the means of data interaction, resulting in poor network latency and stability. Data interaction often takes hundreds of milliseconds.

The characteristic of the LoRaWAN network is determined by its Internet of Things attributes, which enable stable access of more nodes through lightweight management and implementation. The LoRaWAN network achieves its Internet of Things attributes through an adaptive speed, multi-channel, and same-frequency network planning solution.

3. Protocol Standards

The LoRaWAN network is a set of communication protocols defined on the LoRa wireless modulation technology, similar to the TCP/IP protocol. The architecture diagram of the LoRaWAN protocol is divided into several layers, as shown in Figure 2.2.11.

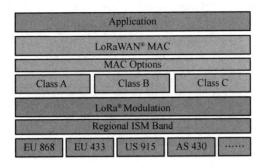

Figure 2.2.11　LoRaWAN Protocol Architecture

① The Application layer provides data for users and is mainly developed by users themselves according to their needs.

② The MAC layer is the link layer protocol, and more detailed information can be found in the LoRaWAN standard.

③ The LoRa Modulation corresponds to the physical layer protocol of OSI, which is actually the wireless modulation technology implemented internally in the sx127x or sx126x baseband chips in specific applications.

④ The last part is because LoRa is a wireless modulation technology, and the transmission of signals relies on radio waves. However, radio waves are actually a public resource. In theory, as long as you have a radio transceiver, you can send radio signals through it. If everyone sends radio signals to the outside world without forming a unified standard, the use of radio will be very chaotic. Therefore, based on the different requirements of different regions, the International Telecommunication Union (ITU) plans different frequency bands for different purposes according to the characteristics of each region, allowing the use of radio to be standardized. For example, China's radio management department has allocated the frequency bands 144～148 MHz and 430～440 MHz to amateur radio services, allowing amateur radio enthusiasts to engage in technical exchanges.

LoRa network divides terminal devices into three categories, specifically Class A, Class B, and Class C.

Class A: Bidirectional communication terminal devices. This type of terminal device allows bidirectional communication, with two downlink reception windows accompanying each uplink transmission. The transmission time slots of terminal devices are based on their own communication needs, with fine-tuning based on the ALOHA protocol. Class A devices have the lowest power consumption, and base station downlink communication can only occur after terminal uplink communication, as shown in Figure 2.2.12.

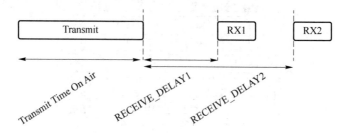

Figure 2.2.12 Class A Behavior Pattern

Class B: Bidirectional communication terminal devices with preset reception time slots. This type of terminal device opens additional reception windows during preset times. To achieve this, the terminal device synchronizes its time with the gateway by receiving a Beacon. Class B terminals allow the base station to know that the terminal is receiving data.

Class C: Bidirectional communication terminal device with the largest receive window. This type of terminal device keeps the receive window open continuously and only closes it during transmission. Class C devices have the longest receive window and consume the most power.

4. Market Applications

In the field of smart agriculture, it is necessary to understand the growth of crops. At this time, sensors such as temperature and humidity, carbon dioxide, and salinity are needed to monitor the growth environment of crops and agricultural products, understand irrigation and growth conditions of crops, and reduce the consumption of water and light resources. Many remote farms and pastures do not have coverage of cellular networks, let alone 4G/LTE. At this time, using LoRaWAN technology to build a private IoT can regularly upload crop data for remote management.

In terms of factory management, signal coverage is achieved using gateways or base stations of LoRaWAN, and LoRaWAN modules are installed at various data collection nodes in the factory. This allows for dynamic collection and transmission of data during production. The data can then be analyzed and optimized through cloud platforms or private platforms, and presented to production management personnel.

For the construction industry, data on safety, environment, processes, and object status are all very important. For example, by installing LoRaWAN smartwatches produced by the Internet of Things, workers can receive instructions and transmit data. This enables better management of construction sites. Smoke sensors and dust detectors can be used to better control the construction environment. Additionally, sensors such as vibration strings can be used for building management.

The logistics industry involves a wide range of areas, and the preferred consideration for the network is low investment and long working life. In order to track pallets and determine the location and status of goods, the facilities throughout the logistics process need to be covered by the network. Therefore, it is required that the network nodes are economically feasible for large-scale deployment and have mobility to be installed on transportation vehicles as a mobile gateway. LoRaWAN's low cost, long battery life, high mobility, and stability in communication during high-speed movement can meet these requirements.

In terms of fire protection, the dynamic guidance system can actively send signals to the dynamic guidance light board inside the building for real-time fire escape instructions. Since LORA operates in the low-frequency band below 1 GHz, there is no need to worry about signal interference from other wireless communications.

In the aspect of smart cities, smart street lights can use LoRa to transmit information about the usage of lamps and bulbs, and promptly notify maintenance units for maintenance and replacement.

2.3 私有协议

2.3.1 华为星闪技术

1. 发展背景及现状

星闪,官方英文名叫作 NearLink,是一种专门用于短距离数据传输的新型无线通信技术。其对标的是蓝牙和 Wi-Fi 等传统技术,并进行了大量的创新升级,实现了速率、时延、传输距离、安全性、可靠性等方面的全面提升,也可以看作是一种增强版的"Wi-Fi+蓝牙"混合体。

2019年5月15日,美国商务部把华为列入"实体清单",对其进行制裁和封锁。随后,蓝牙技术联盟、Wi-Fi 联盟、SD 联盟、国际固态技术联盟、USB 联盟等在内的众多国际技术组织,几乎同时宣布暂停了华为的成员资格(从官网删除了华为的名字,还取消了对华为的技术授权)。在这种情况之下,华为推出了一项超级蓝牙(X-BT)技术。该项技术除涵盖传统蓝牙的所有功能外,还具有电磁环境、感知功率、自适应等功能,综合性能远超传统蓝牙。2019年5月21日,华为正式公布了这项技术,昵称"绿牙"。在发布会的前一天,Wi-Fi联盟、蓝牙联盟等行业组织,又恢复了华为的成员资格。

经过此次除名事件之后,华为决定继续超级蓝牙项目。为了使其能够持续且长久地发展下去,华为号召国内的产业链企业共同加入,组成一个新的技术联盟,并得到了政府层面的大力支持。2020年,在工信部的牵头下,星闪技术标准化正式启动。

2020年9月22日,星闪联盟正式成立。星闪联盟是致力于全球化的产业联盟,目标是推动星闪无线短距通信技术(SparkLink 1.0)的创新和产业生态,使其能够承载智能汽车、智能家居、智能终端和智能制造等快速发展的新场景应用,满足极致性能需求。

2022年11月,星闪联盟正式发布了星闪1.0版本;2023年4月,发布了星闪1.1版本;现在,正在加紧2.0版本的标准化工作。

2. 技术原理

星闪技术的系统框架如图2.3.1所示。

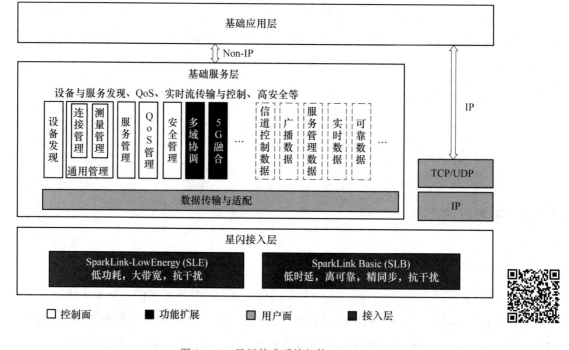

图2.3.1 星闪技术系统架构　　　　彩图2.3.1

星闪无线短距通信技术(SparkLink 1.0)版本的系统一共分为三层,从上到下分别是:基础应用层、基础服务层、星闪接入层。基础应用层用于实现各类应用功能,服务于汽车、家居、影音等不同场景;基础服务层包括很多基础功能单元,通过它们实现对于上层应用功能以及系统管理维护的支持；星闪接入层最为特别，它提供了SLB(基础接入)和SLE(低功耗接入)两种通信接口，分别对应于Wi-Fi和蓝牙两种不同类型的网络场景需求。

(1) SLB(基础接入)

SLB对标Wi-Fi，追求大带宽、大容量、高精度。其支持单载波/多载波，工作频段为5 GHz的非授权频段，最小单载波带宽为20 MHz，向上支持40 MHz/60 MHz/80 MHz/100 MHz/160 MHz/320 MHz，调制方式支持QPSK、16QAM、64QAM、256QAM、1024QAM。

SLB采用超短帧、多点同步、双向认证、快速干扰协调、双向认证加密、跨层调度优化等多项技术，融合了大量5G技术，包括支持类似5G的异步HARQ(混合自动重传请求)技术，类似5G的频域调度设计和帧结构等，以此提升通信性能。

(2) SLE(低功耗接入)

SLE对标蓝牙，追求低功耗、低时延、高可靠性。SLE使用单载波传输，工作频段在2.4 GHz的非授权频段，带宽支持1 MHz、2 MHz和4 MHz，调制方式支持GFSK、BPSK、QPSK和8PSK。SLE支持一对多可靠组播，支持4 kHz短时延，具有安全配对、隐私保护等特性，在尽可能保证传输效率的同时，充分考虑了节能因素。

在组网上，星闪和传统技术类似，也是点对点(P2P)或点对多点(P2MP)。星闪节点分为管理节点(G节点，Grant)和终端节点(T节点，Terminal)。其中，G节点为其覆盖下的T节点提供连接管理、资源分配、信息安全等接入层服务。单个G节点及与其连接的T节点，共同组成一个通信域。

星闪采用了多种安全机制来保护安全和隐私。例如，星闪使用了128 bit的AES加密算法，对数据进行加密和解密，防止数据被窃取或篡改。再例如，星闪使用了双向认证和密钥协商机制，确保设备之间的身份验证和密钥生成。星闪的各方面能力均明显超过了Wi-Fi和蓝牙。

3. 协议标准

星闪在空口技术上采用了很多类似5G的技术，而在技术细节上，其比蓝牙和Wi-Fi更先进。例如，星闪在OFDM和CP设计、时域和频域调度颗粒方面就和5G很类似，甚至用上了5G的Polar编码技术，提升了传输带宽及抗干扰能力。

星闪系统协议栈如图2.3.2所示，分为应用层(OSI 5~7层)、网络与传输层(OSI 3~4层)和接入层(OSI 1~2层)。接入层分为物理层和数据链路层，保障数据的可靠传输。数据链路层包含链路控制层和媒体接入层，这和蓝牙的链路控制层非常相似；而物理层实现了比特流传输功能，星闪的技术要点集中在物理层的提升。

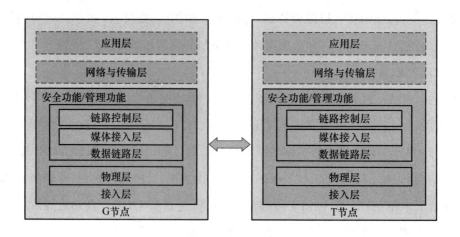

图 2.3.2 星闪系统协议栈

星闪从物理层的频段与传输波形、帧结构、多域协同、低功耗设计、混合自动重传机制、信道编码与调制几个方面对性能提升进行研究。

系统采用 CP-OFDM 波形传输,物理层时间度量为基本时间单位 T_s 的倍数。T_s 定义为 $T_s=1/f_s$,$f_s=30.72\,\mathrm{MHz}$,子载波间隔 $\Delta f=480\,\mathrm{kHz}$。星闪系统的最小载波带宽为 20 MHz,向上支持 40 MHz/60 MHz/80 MHz/100 MHz/160 MHz/320 MHz 规格的载波带宽,分别由连续多个 20 MHz 的载波聚合方式组成。20 MHz 载波由连续 39 个子载波组成,子载波间隔 480 kHz。SLB 使用 CP-OFDM 波形传输,OFDM 会使用全频带的载波进行传输,一定程度上传输速率与频带宽度成正比。

SLE 使用单载波传输,带宽支持 1 MHz、2 MHz 和 4 MHz,调制方式支持 GFSK、BPSK、QPSK 和 8PSK。通过采用 Polar 信道编码提升传输可靠性,减少重传节省功耗,精简广播信道功能和业务以降低拥塞的可能性。

星闪系统的帧结构,从数据构成来说,它和蓝牙的数据帧构成很相似,分为地址、链路层包头、数据。不同的点在于,蓝牙的包头会更冗杂,它包括前导码、MAC 地址、PDU、循环冗余校验等。另外,对于递延时高并发场景,星闪支持初步验证后直接透传,即不传包头直接传数据。

星闪系统采用 TDD(Time-Division Duplex)方式,超帧结构如图 2.3.3 所示,每个超帧包含 48 个无线帧,每个超帧的持续时间为 1 ms,每个无线帧的持续时间为 20.833 μs。

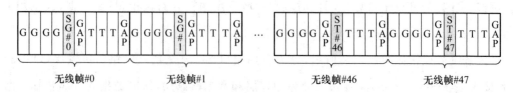

图 2.3.3 星闪系统的超帧结构

传输的每个短帧包都是双工的,并且 G 和 T 的配比是可以调节的,中间有符号位和 GAP 位。每个帧更"轻量化",打包拆包时间更短,帧间隔更短,延时也就更低。

星闪系统使用短帧结构,各设备之间可以快速实现同步。星闪系统通过多 G 节点间的时间/频率同步降低多域间的干扰。同步的过程会采用一个包含同步信号的帧,而各设备间建立同步能有效减少干扰。星闪系统采用 OFDM 波形,在存在多个通信域的场景下,即使不同的通信域使用不同的频点,如果频率差不是子载波间隔 SCS=480 kHz 的整数倍或者定时差异超过 CP,那么就会造成子载波间的干扰。特别的,在干扰来自多个通信域的情况下,以及干扰源比信号源距离接收设备近得多的情况,G 节点之间的时频不对齐引入的干扰会显著降低接收信噪比。

星闪系统支持混合自动重传请求(Hybrid ARQ,HARQ),这是一种 FEC 和 ARQ 相结合的技术,目的是增加链路的传输可靠性。传统 ARQ,当接收端检测到接收的信息中有错误时,接收的错误包信息将被直接丢弃,并请求发送端重传相应的数据包。与传统 ARQ 相比,HARQ 对传统 ARQ 进行了相应的增强,即接收的错误包信息并不会被丢弃,而是会被与重传包信息进行合并,提高接收可靠性。

星闪系统采用基于 Polar 码的异步 HARQ 技术,支持最大 4 个 HARQ 进程,支持 CC-HARQ 方案和 IR-HARQ 方案。CC-HARQ 方案的收益来源于接收端多次软信息合并,提升接收端信息的等效 SNR,降低错误概率;IR-HARQ 方案,根据 Polar 码的特点,重传时扩展母码长度或者发送第一次传输时没有发送的编码比特,在获取能量增益的基础上进一步获取编码增益。

Polar 码是基于信道极化理论构造的一种信道编码,是经过理论分析论证可以达到香农极限的信道编码,可以较好地对抗随机错误。RS 码是一种线性分组码,是基于伽罗华域构建的多进制信道编码,每个符号可包含多个比特,抗突发干扰性能好,可以较好地对抗连续错误。星闪系统使用 Polar 码或 RS 码传输超低时延的小包业务(如车载主动降噪),确保系统可以在不同应用场景下实现高可靠传输。

4. 市场应用

星闪是短距离无线通信技术。它的应用场景,仍然对标 Wi-Fi 和蓝牙。相较于 Wi-Fi 和蓝牙,它的技术性能有了很大提升,可以给用户提供更好的使用体验,也可以进行应用场景拓展。

星闪标准确定后,率先进行商业落地的就是消费电子场景,例如,手机、平板、鼠标、键盘、耳机、音箱、手写笔等。

由于星闪技术的低功耗特点,故其可以大幅降低设备的耗电量,延长待机时长,避免频繁充电。星闪的高速率、低时延,让高品质多声道无损音频传输成为可能。目前的蓝牙,只能支持立体声高清。基于星闪技术的无线鼠标连接的回报率有大幅提升,之前展示的星闪技术鼠标的平均刷新率可以达到 4 kHz,是传统 2.4 GHz 鼠标的 4 倍,平均传输时延是 413.14 μs,是传统 2.4 GHz 鼠标的 1/4。对于游戏玩家来说,这可以显著改善他们的使用体验。

星闪具有极强的抗干扰性。在地铁、高铁、机场等复杂电磁环境场所,它可以更加稳定地工作,减少传输抖动甚至掉线。

(1) 无线投屏和多屏协同

很多用户都喜欢使用无线投屏和多屏协同,这就涉及视频信号的无线传输。利用星闪技术的高速率优势,视频信号可以轻松完成 4K 分辨率的远程投屏;进行多屏协同时,凭借星闪微秒级的时延,可以快速识别焦点设备,做到无缝无感体验。

(2) 智能家居

星闪的连接数达到百级,对于现在越来越多的智能家居终端来说,是很有必要的。用户可以连接更多的设备,也不用担心设备之间的相互干扰。星闪的通信距离长,是蓝牙的 2 倍,有利于实现全屋的无死角覆盖。现在很多家庭有客厅娱乐、体感游戏。星闪可以同时支持多个手柄控制器和体感传感器,让更多的家庭成员参与,也会带来更完美的游戏体验。

(3) 工业制造

现在是万物互联时代,除消费互联网外,行业互联网也发展迅速。我们一直在说 5G 的行业赋能,其实,蓝牙、Wi-Fi 这样的短距无线技术,在企业和园区也有大量使用。星闪的性能比蓝牙、Wi-Fi 更强,在带宽、时延、连接数、安全性、可靠性方面有显著优势,更适合垂直行业场景的需求。

(4) 智能汽车

星闪项目刚启动之时,就已经有车企加入。这说明,汽车内部通信对无线技术具有强烈需求。车机互联、车内影音、雷达传感等,都需要可靠的无线技术进行支撑。

2.3.2 CLAA

1. 发展背景及现状

CLAA 的发展可以追溯到 2016 年,当时 LoRa 技术在物联网领域开始得到广泛应用。LoRa 是一种具有低功耗、长距离传输和网络拓扑灵活等特性的低功耗广域网(LPWAN)协议,适用于各种物联网应用场景。然而,在中国 LoRa 技术的应用和发展面临着一些挑战,如缺乏统一的标准和规范,存在不同厂商之间的兼容性问题等。

2019 年,中兴通讯发起了开放、共享的物联网生态圈,CLAA 是在这样的背景下应运而生的。它致力于推动 LoRa 技术在中国的应用和发展,自 2016 年 1 月 CLAA 成立以来,得到了物联网 LPWAN 业界的广泛关注和支持。如今,CLAA 已经发展了芯片、模块、终端、垂直应用整个 LPWAN 产业链的 1 300 多家正式成员。通过合作,CLAA 已经形成了从 LoRa 芯片、模组、终端到系统集成商、解决方案提供商等各环节的完整产业链,成员企业涵盖了芯片模块、终端传感和系统集成商等多个领域,如克拉科技、Semtech、利尔达、唯传科技等。CLAA 的 LPWAN 技术在智能抄表、智能停车、智能照明、智慧农业等各个领域得到了广泛应用。同时,CLAA 也积极推动不同行业之间的合作和交流,促进跨行业物联网应用的发展。CLAA 不仅致力于解决当前物联网应用中面临的挑战,还不断进行技术创新和模式创新,推动物联网技术的进步和发展。例如,CLAA 推出了全球首个 5G LPWAN 标准,并积极探索与其他技术的融合和应用。CLAA 不仅在中国得到广泛应用和支持,还积极拓展海外市场。目前,CLAA 已经在美国、欧洲等地建立了分支机构,推动中国 LoRa 技术和标准走向世界。

CLAA 作为中国 LoRa 技术的引领者和推动者,通过打造完整的产业链、推动 LoRa 技术的广泛应用、坚持持续创新以及推动国际化发展等举措,为 LoRa 技术在中国的应用和发展做出了不可忽视的重要贡献。

2. 技术原理

LoRa 网络的结构主要由以下几个部分组成。

① 终端(内置 LoRa 模块):设备的末端,利用 LoRa 协议与网关进行通信。

② 网关(或称基站):连接终端设备和后端中央服务器的中继。LoRa 网关采用星形拓扑结构,对从终端设备接收到的数据进行处理和转发。

③ 网络服务器:整个 LoRa 网络的核心,处理和存储从网关发送来的数据,并对其进行相应的处理。

④ 应用服务器:用来处理特定应用的数据和服务。

LoRaWAN 网络架构的独特之处在于它的星形拓扑结构,这种结构使得终端设备能够通过网关与中央服务器进行通信。而为了实现这一目标,所有的节点(终端设备和网关)都必须遵守 LoRa 联盟规定的 MAC 层通信协议,这样不同硬件厂商的设备才能互相接入。

在 LoRaWAN 的基础上,CLAA 参考电信运营商网络标准和相关协议,定义了基于 LoRa 技术的运营级物联网网络标准和网络协议。

CLAA 网络将应用、运营、网络、终端进行解耦,实现专业分工。这种分层解耦的设计使得用户可以像购置移动蜂窝网终端(如手机)一样,随时随地部署传感终端,采集所需数据。同时,传感和应用厂商可以专注终端和应用开发,实现规模发展,更能发挥各自优势。

CLAA 物联网的网络实现充分采用当前先进的云计算技术,以及弹性云端、全网服务的方案。CLAA 物联网的云端服务网络功能(核心网)部署在公用或专有云端服务器上,所有 CLAA 网络节点(CLAA 基站)和应用业务均可通过互联网或行业专网与云端服务器连接,组成统一的物联网网络。这种弹性云端的部署方式可根据物联网基站部署数量、传感终端接入数量、业务应用系统接入数量等进行弹性扩充,实现了物联网的快速部署商用和灵活扩展演进。

同时,CLAA 物联网支持独立式云化应用,使得传统物联网接入和应用不再局限于独立项目模式和限定场所、场景。通过 CLAA 物联网,人们可以实现对专业行业性的大范围管理,如物流、消防部门、生产安全管理、物业管理等。这些行业的数据采集和监控可以更加便捷地实现云化,提高管理效率和响应速度。

在网络安全和管理方面,CLAA 提供了包括链路加密、自部署注册服务器、终端采用安全芯片等安全措施。同时,CLAA 核心网内置注册服务器,接口开放,方便用户进行远程管理和维护。这些安全措施和管理功能为物联网的安全、稳定运行提供了有力保障。

3. 协议标准

CLAA 协议标准的具体内容涉及多个方面。

首先,网络架构基于 LoRaWAN 协议,由终端设备、网关和网络服务器组成。其次,协议规范包括数据传输协议、设备管理协议、网络安全协议等,它定义了设备之间的通信方式、数据格式、消息路由、安全策略等。再次,硬件要求包括终端设备和网关的硬件规格、接口定义、电源要求等,以确保设备之间的兼容性和稳定性。应用场景方面,CLAA 的协议标准定义了相应的数据格式、传输速率、工作模式等,以适应不同的应用需求。最后,CLAA 致力于推动不同厂商之间的互联互通,制定了 LoRaWAN 协议的统一规范,以实现不同设备之间的互操作性和兼容性。

具体来说,CLAA 的协议标准包括以下几个方面。

(1) 网关规范

CLAA 的网关规范包括网关的硬件规格、接口定义、电源要求等,以确保网关能够与终端设备进行稳定可靠的通信。

(2) 数据传输协议

CLAA 的数据传输协议基于 LoRaWAN 协议,定义了数据传输的方式、数据格式、消息路由等。此外,CLAA 还优化了协议,以提高数据传输的效率和稳定性。

(3) 设备管理协议

CLAA 的设备管理协议定义了设备注册、设备配置、设备状态监测等功能,以确保设备能够正常运行并满足应用需求。

(4) 网络安全协议

CLAA 的网络安全协议包括加密算法、密钥管理、安全认证等,以确保数据的安全性和隐私性。

(5) 应用场景规范

针对不同的应用场景,CLAA 的协议标准定义了相应的数据格式、传输速率、工作模式等,以满足不同应用的需求。

(6) 互联互通规范

CLAA 致力于推动不同厂商之间的互联互通,制定了 LoRaWAN 协议的统一规范,以实现不同设备之间的互操作性和兼容性。

CLAA 在 LoRaWAN 的基础上优化了协议,如图 2.3.4 所示,构建了共建共享的 LoRa 应用平台。CLAA 提供网关和云化核心网服务,可快速搭建起 LoRa 网络的物联网系统的应用。

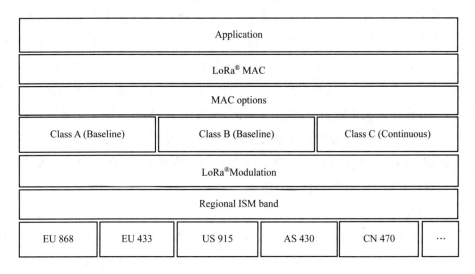

图 2.3.4　CLAA 协议架构

在中国区,LoRaWAN 使用的信道频率为 CN779、CN470 等;使用的调制模式为 LoRaWAN 调制模式;支持的速率包括 0.3~5 kbit/s 等;支持的功率符合当地的法规要求,但具体数值需要根据实际应用场景和法规来确定;支持的空速包括 SF7-SF12 等;最大负载大小需要符合当地的法规要求和实际应用场景。除了以上几个方面,CLAA 的协议标准还包括一些其他参数,如网关地址、数据包格式等,这些参数也需要根据实际应用场景来确定。

需要注意的是,CLAA 的协议标准是在不断更新和演进的,具体的参数和要求可能会随之变化。因此,在实际应用中,厂商需要根据具体情况来确定使用哪些参数和要求。通过制定统一的规范和标准,CLAA 有助于促进不同厂商之间的合作和竞争,共同推动 LoRa 技术在更多领域的应用和发展。

4. 市场应用

(1) 智慧水务

智慧水务是基于物联网理念的信息化解决方案,其核心就是把传感器嵌入或装备到水源、供水系统、排污系统中,并且被普遍连接,形成"感知物联网"。然后利用云计算、大数据、人工智能等技术,通过全面感知、无线通信、自主诊断、场景应用等方面的结合,构建成全方位的城市水体(包括河湖,污水排水管网等)生命周期管理系统,实现政府管理机构、企业和社区与水务管理系统的整合。克拉科技智慧水务解决方案选用可规模化部署、低功耗、低成本、高稳定性的一系列智能监测设备,通过CLAA无线网络,在无需布线的情况下,可快速实现物联网系统的部署,实现水表远程集抄、供水管道压力液位监测、井盖安全监测、河流河道水质监测、水文雨情监测、水泵站监测等智慧化应用。

(2) 智慧农业

党的十九大提出了"加快推进农业农村现代化",推动工业化、信息化、城镇化、农业现代化同步发展。各级人民政府出台相关文件,加快推进农业现代化建设,推进农业现代化与信息化同步发展。利用物联网、大数据、云计算等新一代信息技术整合社区资源,依托一张统一标准的,覆盖农业大田、大棚的LoRa物联网络,以及在农业领域内的各类终端传感设备,如空气温湿度计、土壤温湿度计、太阳辐射监测、环境监测、水质监测、土壤pH值、土壤盐度设备等,为农户提供高效、便携和智慧的服务。智慧农业为新农人运营及工作人员提供包括规划部署、日常运维、实时感知的设备全生命周期管理;同时整网可支持快速的功能扩展并与第三方系统对接,让农业网络更好地为运营方提供业务服务。

(3) 智慧市政

截至2019年10月,加入CLAA联盟的各类厂商企业已有1 200多家。CLAA专注智慧市政领域,将为客户提供集成、统一、融合、一站式解决方案。通过部署CLAA网关,将各种感知终端上的数据收集汇总上来,以监测周围的空气质量、气象指标、环境指标是否达标,重要的市政基础设施,如井盖、垃圾桶、路灯、环卫车辆、雨污管网、雨水管道等是否完好正常,并开展智慧停车、智慧园林等民生应用。智慧市政集成开发多功能业务平台,通过数据挖掘、机器学习等大数据分析,为各个职能部门的决策服务和政策制定提供响应和支撑,对接好市政的各个职能部门,实行处理机制的快速响应,为市民提供舒适、安全、便捷、低碳和高效的生活工作环境。

(4) 智慧电力

中兴克拉科技端边管云一体化低功耗物联网解决方案,利用物联网、大数据、云计算等新一代信息技术深入电力物联网内外部应用,与新一代电力系统相互渗透和深度融合,实时在线连接能源电力生产和消费各环节的人、机、物。对内承接输电业务、变电业务、配电业务、用电业务、经营管理等五大业务场景的智能物联设备的本地接入,对外拓展综合能源服务、美丽乡村等新兴业务,改变以往单一的业务模式,促进新兴业务和电网业务互利共生、协同发展。

(5) 城市安全生命线

以物联网、云计算、大数据、BIM/GIS等技术为支撑,CLAA动态感知城市生命线,事件智慧预警,有效协同处置保障城市生命线安全。基于CLAA,人们打造"陆海空天地"一体化的城市安全风险感知体系;重点提升燃气、给排水、电梯、综合管廊、电力管网和综合交通等专项领域的监测预警能力;完善城市生命线工程基础平台建设,拓展生命线周边环境风险感知能力、通信保障能力、防灾韧性等城市生命线工程安全建设相关能力;加强供水、排水、燃气、桥梁涵洞、综合管廊等城市生命线泛场景数据的获取、分析和研判,实现数字化、精细化、可视化管控。让安全管控实现从"以治为主""被动应付"到"以防为主""主动监管"的科学转变。从而达到"提升管控水平,提高安全系数,降低安全监管强度,实现降本增效"之目的。

2.3.3 Mesh Network

1. Background and Current State of Development

Wireless Mesh Network, also known as a multihop network, is a new type of wireless network technology. It is composed of multiple mobile terminals with wireless transmission and reception capabilities, forming a multihop, self-organizing, self-repairing, and decentralized network. It is a type of mobile computer communication network. Compared to traditional wireless networks, wireless mesh networks have higher reliability, stronger mobility, and faster deployment flexibility. Without a central control node, it can quickly establish a wireless communication network through the cooperation and route selection between node devices, without the need for any other fixed communication network infrastructure, at any time and place.

The concept of wireless mesh networking was first proposed in the mid-1990s and originated from military communication needs. In 1972, the United States DARPA (Defense Advanced Research Projects Agency) launched the Packet Radio Network (PRNET) project, which mainly focused on the data communication applications of packet radio networks in battlefield environments. After completing this project, DARPA initiated the Survivable Adaptive Network (SURAN) project in 1993, aiming to further expand the achievements of PRNET to support larger-scale networks and develop new adaptive network protocols to adapt to rapidly changing battlefield environments. In order to develop a mobile information system that meets military application needs, can be rapidly deployed, and has high resistance to destruction, DARPA launched the Global Mobile Information Systems (GloMo) project in 1994. This project conducted comprehensive and in-depth research based on existing achievements and has been ongoing until now.

The IEEE 802.11 standard committee, established in 1991, adopted the term "Ad hoc network" to describe this special type of peer-to-peer, multihop, mobile communication network. The civilian mobile communication networks we are familiar with are usually centralized and rely on pre-existing network infrastructure to operate. For example, cellular mobile communication systems require support from base stations, and wireless local area networks typically operate in a mode with Access Points (APs) and wired backbone networks. However, centralized mobile communication networks may not be suitable for certain special occasions, such as rapid advancement and deployment in battlefield environments or post-natural disaster rescue operations. In situations where there is no pre-existing network infrastructure or the pre-existing network infrastructure has been destroyed, mobile communication networks need to have the ability to quickly and automatically form temporary networks. Ad hoc networks can meet these requirements.

For the civilian mobile communication field, there is also an urgent need for a network similar to one with Ad hoc capability. Ad hoc network research is based on military needs and requires high costs, so Ad hoc networks cannot be directly applied to civilian mobile communication. In this scenario, wireless mesh networks were born, which research wireless multihop network technology that is more suitable for civilian mobile communication.

Wireless mesh networks are the result of the combination of wireless local area networks and mobile ad hoc networks, and it is a new network architecture. Wireless mesh technology has also gradually gained attention. In 2000, Mesh Networks, a company in the United States, acquired patents from ITT company, mainly involving some patents of the US military tactical mobile communication system. They transformed it into civilian technology and launched a series of civilian wireless multihop self-organizing network products, which gained wide market recognition. During the same period, Nokia, Tropos, Nortel Networks, SkyPilot, Radiant, and other companies also jointly developed related wireless mesh network products. In 2005, Motorola saw the potential and acquired mesh networks company. In addition, during the research process of the IEEE 802.16 wireless metropolitan area network standard, support for the wireless mesh network structure was added, and related protocol standards were introduced. Wireless mesh networks also entered a period of rapid development.

Wireless mesh networks have a wide range of applications, including wired/wireless access, wireless transmission, and simple networking. Its main application scenarios include wireless Internet access for home users, business users, enterprises, schools, and even entire cities, enabling Internal networking or wireless Internet access. Wireless mesh networks have enormous market potential.

2. Technical Principles

Wireless mesh networks inherit the characteristics of Ad hoc networks, such as decentralization, lack of infrastructure, multihop, and self-organization, and develop a new networking architecture on this basis to meet their main purpose of providing IP broadband access.

In traditional wireless access technologies, the main topology used is point-to-point or point-to-multipoint. Such topologies generally have a central node, such as a base station in mobile communication systems or an AP in IEEE 802.11 WLAN. The main role of the central node is to connect wireless terminals through single-hop wireless links and control the wireless terminals' access to the wireless network. The central node is connected to the wired backbone network through wired links.

Wireless mesh networking, using a mesh topology, is a multipoint-to-multipoint network topology. Each network node is connected to other network nodes through wireless multihop. The types of wireless mesh network nodes include wireless mesh client nodes and wireless mesh router nodes.

In addition to the gateway/relay function of traditional wireless routers, wireless mesh routers also support routing functions for mesh network interconnection. mesh routers typically have multiple wireless interfaces, which can be built based on the same wireless access technology or different wireless access technologies. Compared with traditional wireless routers, wireless mesh routers can achieve the same wireless coverage range with lower transmission power through wireless multihop communication. Wireless mesh terminals also have certain mesh network interconnection and packet forwarding functions, but generally do not have gateway bridging functions. Usually, Mesh terminals only have one wireless interface, with much lower complexity compared to mesh routers.

According to the different functions of each node, the wireless mesh network structure can be divided into three types: client mesh structure (flat structure), backbone mesh structure (hierarchical structure), and hybrid structure.

The client mesh structure is composed of mesh clients, as shown in Figure 2.3.5, providing point-to-point wireless services between user devices. The clients form a network that can provide routing and configuration functions, supporting user terminal applications. Since the nodes in the network do not need to have gateway or relay functions, mesh routers are not required. In this network structure, clients typically use only one wireless technology. Data packets sent by any node can be forwarded through multiple nodes to reach the destination node. Although nodes do not need gateway and relay functions, routing and self-organization capabilities are necessary.

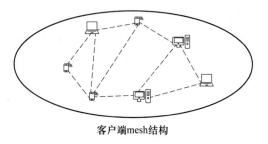

客户端mesh结构

Figure 2.3.5 Client Mesh Structure

The backbone network of the mesh structure, as shown in Figure 2.3.6, is formed by the interconnection of mesh routers, and the wireless mesh backbone network is connected to the external network through these mesh routers. In addition to the gateway and relay functions of traditional wireless routers, mesh routers also have routing functions that support mesh network interconnection. They can achieve the same wireless coverage range with much lower transmission power through wireless multihop communication.

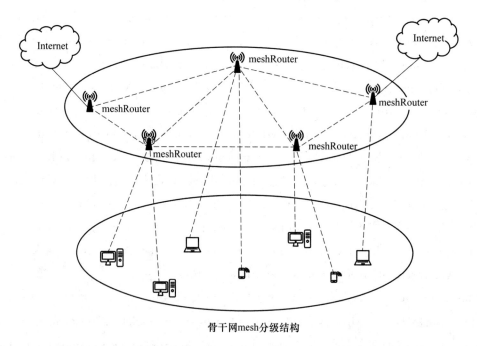

骨干网mesh分级结构

Figure 2.3.6 Backbone Network Mesh Hierarchical Structure

The hybrid architecture is a combination of these two architectures, as shown in Figure 2.3.7. Mesh clients can access the network through mesh routers and form a mesh network with other mesh clients. The backbone network consisting of mesh routers, provides connectivity to other network structures, such as Internet, WLAN, WiMax, cellular networks, and sensor networks. At the same time, in this wireless mesh network structure, the routing function of the clients is enhanced, and the network used for the wireless mesh network can also expand the network coverage.

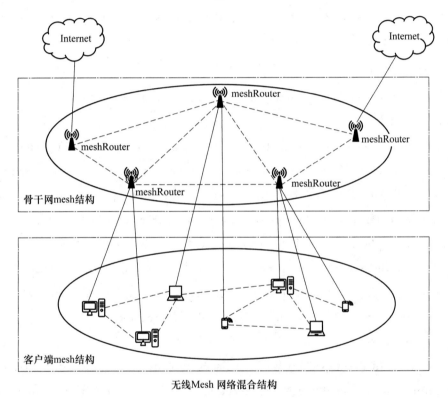

Figure 2.3.7 Hybrid Architecture of Wireless Mesh Netowrks

The main characteristics of mesh networks include：

Decentralization：Mesh networks do not have a central node. Each node is equal and can act as a relay to forward data, providing high fault tolerance and reliability.

High flexibility：Nodes in a mesh network can dynamically join or leave the network, and the network topology can change at any time, providing high flexibility and scalability.

Strong scalability：Mesh networks can expand their coverage and capacity by adding more nodes, offering high scalability.

High reliability：Nodes in a mesh network can transmit data through multiple paths, increasing the reliability of the network.

Fast transmission speed: Mesh networks use multihop communication technology, where each node can transmit data to multiple nodes, thereby improving transmission speed and throughput.

Mesh networks typically use wireless technologies, such as Wi-Fi, Bluetooth, Zigbee, etc., and can be applied in fields such as the Internet of Things, smart homes, and city management. Mesh networks can automatically discover and configure networks, enabling fast deployment and easy installation, making them important for the future development of the Internet of Things.

3. Protocol Standards

In wireless mesh networks, there are several common mesh protocols.

(1) Wi-Fi Mesh

Wi-Fi Mesh is a wireless mesh network protocol based on the 802.11 standard, which enables communication and interconnection between terminal devices using Wi-Fi technology. Wi-Fi Mesh has high data transmission rates and large coverage areas, making it suitable for large enterprises and campuses. In 2011, IEEE established the 802.11s standard as a supplement to the 802.11 MAC layer protocol, which specifies how to build a mesh network based on the 802.11a/b/g/n protocols (at that time, 802.11n was the latest Wi-Fi standard). The standard includes mesh frame formats, extended mesh frame headers, mesh protocol procedures, mesh security mechanism and mesh routing.

(2) Easy-mesh

A basic standard developed by Wi-Fi Alliance for interconnecting different APs. This standard defines the protocol for interconnecting APs from different manufacturers. The EasyMesh protocol is called the Multi-AP Specification within the Wi-Fi Alliance. EasyMesh was released in 2018 with version 1.0 and is currently at version 3.0. In an EasyMesh network, there is a Controller role that manages the entire network, and all Agent role APs are connected to it (including indirectly). The Controller role is usually the main wireless router for the home network, which includes the management process of the Controller and also has an Agent AP. Of course, this Agent AP is also under the management of the Controller, which is more cost-effective. EasyMesh forms a tree-shaped network topology by connecting in a similar way to traditional AP+Client.

(3) Bluetooth Mesh

Bluetooth Mesh is a wireless mesh network protocol based on the 802.15.1 standard. It enables communication and interconnection between terminal devices using Bluetooth technology. Bluetooth Mesh has low power consumption and small device cost, making it suitable for scenarios such as smart homes and the Internet of Things.

(4) Zigbee Mesh

Zigbee Mesh is a wireless mesh network protocol based on the 802.15.4 standard. It enables communication and interconnection between terminal devices using Zigbee technology. Zigbee Mesh has low data transmission rate and low power consumption, making it suitable for scenarios such as smart homes and industrial automation.

(5) 6LoWPAN Mesh

6LoWPAN Mesh is a wireless mesh network protocol based on the 6LoWPAN standard. It enables communication and interconnection between terminal devices using wireless sensor network technology. 6LoWPAN Mesh has low power consumption and small device cost, making it suitable for scenarios such as the Internet of Things and smart cities.

These mesh protocols have slight differences in application scenarios, data transmission rates, coverage ranges, and device costs. However, they all have the basic characteristics and functions of wireless mesh networks, enabling communication and interconnection between terminal devices. In practical applications, suitable mesh protocols can be selected based on specific scenarios and requirements to implement the construction of wireless mesh networks.

4. Market Applications

Compared to traditional infrastructure models, wireless mesh networks have strong scalability, flexibility, and self-organization. Without a central control node, it can achieve wider coverage, larger capacity, and stronger robustness in wireless communication networks through collaboration and route selection between nodes. Its application scope can cover various fields such as industry, commerce, healthcare, home, armed police, fire protection, civil defense emergency, three defense emergency, water conservancy flood control, power inspection, power emergency repair, "Bright Project" railway emergency repair, maritime law enforcement, maritime supervision, customs border defense, port monitoring, forest fire prevention, oilfield theft prevention, military reconnaissance, TV broadcasting, etc. It provides high-quality real-time mobile transmission of images, voice, and data in various complexenvironments such as cities, seas, and mountains.

Smart city is a concept that is constantly evolving. Its goal is to use new generation information and communication technologies to achieve intelligent management and operation of cities, and to achieve sustainable development in terms of economy, society, and ecology, thereby making people's lives in cities more convenient, beautiful, and harmonious. Smart city is an inevitable stage in the process of informatization of cities. From a technical perspective, smart city requires the realization of intelligent perception, ubiquitous interconnection, pervasive computing, and integrated applications represented by mobile Internet, Internet of Things, cloud computing, and big data. Data network is the "nerve" and "blood vessels" of smart cities, carrying high-speed data transmission in future smart cities. Considering the scale of future cities, it can be foreseen that data transmission networks need to have characteristics such as high bandwidth, high reliability, wide coverage, and easy scalability, while also having relatively low wiring costs.

Wireless Mesh Networks (WMNs) are a new generation of large-scale, high-performance wireless access networks that possess the characteristics required by smart cities for data transmission networks. Therefore, mesh networking technology will become one of the key technologies for achieving high-speed data transmission in future smart cities. In the future, wireless mesh networks will be fully integrated with WLAN, NB-IoT, LoRa, 3G/4G/5G, and other networks, comprehensively improving the quality of wireless networks. At the same time, they will be combined with wireless sensors and community networks to form comprehensive networks, becoming part of the Internet of Things, allowing users to fully enjoy the convenience and excitement of wireless networks.

Campus wireless networks have their characteristics: first, they cover a large geographical area with a large number of users and high communication volume; second, high network coverage is required, and the network must be able to achieve seamless roaming between indoor and outdoor areas, auditoriums, dormitories, libraries, and public places; third, load balancing is very important, as communication congestion is prone to occur when students gather in a certain area and use the network simultaneously. Using wireless mesh self-organizing networks, it is easy to adjust the number and position of nodes, achieve network upgrades, and realize seamless roaming indoors and outdoors. Currently, manyuniversities in China, such as Ningxia University, Chengdu University of Information Technology, Beijing University of Posts and Telecommunications, and Beijing Institute of Technology, have already established or are in the process of building wireless mesh networks.

Hospital buildings are densely and complexly constructed, and some areas need to prevent electromagnetic radiation. Wiring is difficult, and the network requires high robustness: for important activities such as surgeries, any network failure can have catastrophic consequences. Using wireless mesh networking can solve these problems. The wireless network topology adjustment is simple, and the robustness and high bandwidth of the network are more suitable for deployment in hospitals and other public places.

Wireless mesh networking is very suitable for places with remote geographical locations, difficult or uneconomical wiring, and the need to provide broadband wireless Internet access to users, such as tourist attractions, resorts, motels, etc. Wireless mesh networking can provide broadband services to these places at the lowest cost.

For places that require quick deployment or temporary installation, such as exhibitions, trade shows, disaster relief, etc., wireless mesh networking is the most cost-effective networking method that can minimize costs.

Wireless mesh networking has significant advantages in network deployment for professional industries such as transportation, banking, electricity, and energy. Taking urban public transportation systems as an example, using mesh systems to build smart wireless networks for buses, stations, and major routes supports smart public transportation applications. This includes monitoring functions such as real-time video transmission, vehicle information monitoring, and passenger flow monitoring; management functions such as vehicle scheduling and dispatching; and information services such as passenger Internet access, in-vehicle content services, and electronic signage.

Mobile wireless mesh communication system, especially suitable for use in construction sites. The monitoring heads (which can be added according to actual needs) are installed at key monitoring locations in the construction site, such as east, south, west, and north, and can also be installed on the cranes in the construction site. In addition to implementing comprehensive monitoring of the construction site, it also allows crane operators to see the position of the lifted objects, enabling more precise operations and ensuring construction safety.

In large ports, we often see gantry cranes. Due to the special nature of gantry cranes' work, such as handling large cargo and containers, only one driver can follow the cargo in the control room and approximate the unloading position, and then ground personnel command through the intercom to achieve the desired position for the cargo. The driver can only operate by listening to the conversation. Without visual implementation, they are helpless in case of sudden emergencies.

The mobile wireless monitoring and security management system is not limited by the mobility of gantry cranes and uses wireless transmission. It is not restricted by traditional network cables and has a longer lifespan. Just connect it to the power supply and fix the high-definition camera in the desired position. The device will automatically detect and operate, and the gantry crane operator can receive real-time high-definition monitoring images through the NVR display in the background, improving blind spot operation.

In the military field, maintaining smooth communication at all times determines the outcome of a battle. Wireless mesh networks have the advantages of automatic networking, self-configuration, and self-repair. There is no single point of failure, and the network can continue to operate even if any node stops working. The entire network has strong vitality and is known as the "indestructible network". Wireless mesh networks are used to integrate various military service resources and assist in unified command and joint operations of the military. They provide temporary networking, communication command, and logistical support wireless network services for military camps, exercises, and wartime needs. They achieve unified command and coordination between different functional units of the military. With the flexible networking capabilities of wireless mesh networks, the military wireless network system can be applied in the following ways.

Mobile Group Network with Military Vehicles: Install wireless mesh nodes on military vehicles to form a mobile mesh network. It is used for group marches, exercises, emergencies, or wartime needs. The mesh nodes on the vehicles can self-configure and automatically establish connections, forming a cellular wireless network that supports wireless data, wireless voice, and wireless video surveillance. In situations where vehicles are moving and formations are constantly changing, the mesh structure can automatically form a network, ensuringuninterrupted connectivity. Nodes can automatically hop connections to maintain the stable operation of the entire network.

第3章 物联网场景实践

3.1 智慧社区

3.1.1 Smart Home

With the development of science and technology and the improvement of people's living standards, individuals are paying more and more attention to the comfort, safety, and convenience of their living environments. In recent years, emerging smart home systems have met this demand.

Smart home is a platform based on residential buildings, which integrates the facilities related to home life through comprehensive wiring technology, network communication technology, security technology, automatic control technology, and audio and video technology. It constructs an efficient management system for residential facilities and household schedules, enhancing home security, convenience, comfort, and artistry, and achieves an environmentally friendly and energy-saving living environment. In other words, smart home is not merely a single product, but a technology that connects all the products in the home into an organic system, allowing the owner to control the system anytime and anywhere.

Due to the many wired communication connections and complex methods, there are many hidden dangers within the lines, and the overall degree of intelligence is not high. Therefore, wireless communication has been the focus of research in various communication industries in recent years. The wireless transmission technologies mainly used in smart home systems are ZigBee, Wi-Fi, Bluetooth, and Z-Wave.

Due to the popularity of Bluetooth, smart home systems controlled by Bluetooth have inherent advantages over other technologies. The low power consumption, fast transmission speeds, and long distance characteristics of Bluetooth technology further enhance its applicability in smart homes. With the further development of wireless Bluetooth technology, the continuous expansion of product scale, the increasing maturity of technology, and the decrease in prices, Bluetooth technology is gradually being applied to the industrial field and households. The special characteristics and advantages of wireless Bluetooth communication make it a flexible and suitable for application in smart home information networks.

With the popularization of network and automation technology, scattered and single-function home appliances are gradually eliminated, and smart home networks and integrated smart homes will undoubtedly become the mainstream of China's future development. From the initial use of Bluetooth technology was utilized for data transmission in mobile phones, leading to the rise of Bluetooth headphones and Bluetooth wireless mice, and now the most popular Bluetooth smart home systems, the convenience brought by Bluetooth to people's lives is self-evident. In the future of smart home, adopting Bluetooth undoubtedly brings great convenience to users. Bluetooth is the optimal choice for solving the needs of modern smart homes.

1. Scenario Analysis

In home life, consumers' demands for smart homes can be divided into two categories: ensuring safety and privacy, and substituting household chores while regulating the environment. From the perspective of functionality and performance, these demands are manifested in the following aspects.

Smart home systems should include the following functions:

(1) Smart security scenarios

In smart security scenarios, functions such as home security monitoring, anti-theft alarms, and access control management are core. Key devices for implementing these functions include smart locks, smart cameras, and smart alarms. For example, when someone breaks into the home, the smart security system will automatically trigger the alarm and send the alarm information to the mobile phones of family members, reminding everyone to take timely measures.

(2) Intelligent Lighting Scenarios

Intelligent lighting scenarios are another important application in smart homes. Through an intelligent lighting system enables the intelligent control and management of home lighting can be achieved. Smart bulbs, smart switches, and other devices are the foundation for implementing this functionality. With an intelligent lighting system, functions such as automatic light adjustment, scheduled switching, and remote control can be realized, makinghome lighting more intelligent and comfortable.

(3) Connected Control Devices

Connected control devices are the core components in smart home scenarios. Through connected control devices, users can achieve unified management and control of various smart devices in their homes. For example, users can utilize their smartphones to control the smart router at home to download movies, or they can use their smartphones to control the rice cooker remotely to start cooking.

(4) Smart Home Appliances

Smart home appliances are an important component of smart home scenarios. Through smart home appliances, users can achieve functions such as remote control, scheduled switching, and voice control of home appliances. For example, users can manage the on/off and temperature of their air conditioner through their mobile phones, and they can also control the play and pause of the TV through voice commands.

(5) Smart Bathroom

Smart bathroom scenarios are another specific application scenario. Through devices such as smart toilets and smart showers, users can enjoy a more intelligent and comfortable bathroom experience. For example, users can control the water temperature, pressure, and duration of the shower through their mobile phones, and they can also control the on/off and flushing function of the toilet through voice commands.

(6) Smart Sleep

The smart sleep scenario represents a specific application scenario in smart homes. Through devices such as smart mattresses and smart pillows, users can monitor their sleep quality and receive personalized recommendations through data analysis. At the same time, these devices can automatically adjust the hardness of the mattress and pillows according to the user's sleep habits, creating a more comfortable sleep environment.

(7) Care for the Young and Elderly

The care of young children and elderly scenario is a specific application scenario that utilizes technology to enhance their well-being. Through devices such as smart cameras and smart speakers, parents or caregivers can monitor the safety and health of children in real-time. Additionally, they can use voice control to play children's stories or music for entertainment. At the same time, these devices can provide personalized care services based on the lifestyle and needs of the elderly.

Smart home systems should possess the following capabilities:

① Stability. The system should possess stable operation capabilities to avoid inconvenience caused by malfunctions.

② Security. The system should possess security measures in place to prevent information leakage and security vulnerabilities.

③ Scalability. The system should possess the ability to expand and meet the constantly changing needs of users.

④ Usability. The system should possess a simple and user-friendly interface and operation methods to ensure ease of use and management for users.

Various smart home scenarios are shown in Figure 3.1.1.

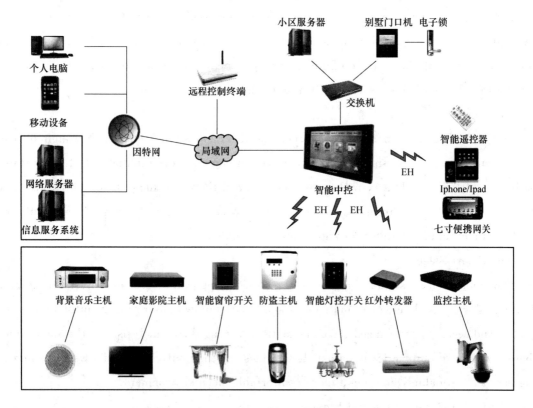

Figure 3.1.1 Smart Home Scene Diagram

Specific requirements and applications need to be clarified, such as:

Home security systems can be divided into three parts: defense alarm systems, video surveillance systems, and access control systems.

Real-time detection and remote monitoring needs of users can be met by using Bluetooth control for defense alarms, thereby reducing false alarm rates.

The home video surveillance system consists of three main parts: the monitoring center, the host control, and data processing. By using the Internet and connecting to the community network, images can be wirelessly transmitted through Bluetooth for USB image acquisition, and motion detection can be performed on the images. SMS alarms can be sent using GPRS.

The access control system mainly includes two main functions: automatic recognition and security management. By detecting the passage of objects through the channel, safe control of entry and exit can be achieved, and real-time monitoring and post-event inspection can be integrated to identify and record personnel access. Bluetooth technology can be used to control access through a mobile client password.

The home appliance system in smart homes generally consists of five parts: smart wearables, lighting control, audio and video entertainment, smart switches, and smart appliances.

Smart wearables refer to independent wearable electronic products such as smartwatches and smartphones. They use low-power Bluetooth communication technology to detect environmental data such as temperature, humidity, and lighting, and use wireless transmission technology for security alarms. The Bluetooth lighting control system uses mobile phones, PDAs, etc. as control terminals to control the on/off, color, and brightness of the lights, and with the help of the timing function, it controls the actions of the lights.

The audiovisual entertainment system is based on the concept of wireless Bluetooth audiovisual stereo walls, integrating TV, Internet, wireless Bluetooth, and stereo sound devices to achieve wireless connectivity for audiovisual entertainment. The smart switch receiver receives instructions transmitted through Bluetooth to control the lighting.

The receiver is connected between smart power facilities and lamp managers, replacing the traditional manual switch mode, allowing for free switching of lamps, light sources, brightness, etc. Smart home appliances include large and small appliances such as refrigerators, washing machines, microwaves, and televisions. They are connected to the main station through Bluetooth wireless communication technology and obtain information about the usage of home appliances through the Internet network.

In the following, we will use an example to illustrate the process of analyzing scenarios and verifying solutions for a smart home system.

Customer requirements.

Requirement 1: The front door should have a fingerprint lock that can register and identify the owner's fingerprint.

Requirement 2: The entrance should have a brightness sensor. When the front door is opened, the hallway lights up, and after passing through the hallway, the lights automatically turn off.

Requirement 3: The living room should have a wireless touch screen that can view the lighting, temperature, and humidity conditions of each room, and control the lighting, air conditioning, and curtains in each room.

Requirement 4: The bathroom should have an infrared sensor. The lights should slowly turn on when entering the bathroom and automatically turn off after a period of time when leaving.

Requirement 5: The kitchen should be equipped with an abnormal alarm device that can automatically sound an alarm when the CO concentration or smoke concentration exceeds the standard.

Requirement 6: After leaving home, it should be possible to remotely view the indoor situation and control the on/off switches of indoor appliances and devices.

2. Project Scheme Design

The design of a smart home project scheme needs to consider multiple aspects, including device selection, system architecture, function design, user experience, etc.

(1) Device Selection

In terms of device selection, it is necessary to choose suitable smart devices based on actual needs. For example, smart door locks, smart cameras, smart lighting, smart home appliances, etc. At the same time, the compatibility and scalability of the devices need to be considered so that it is convenient to upgrade and expand when adding new devices or functions in the future.

(2) System Architecture

In terms of system architecture, it is necessary to design a stable, efficient, and scalable system architecture. Device connectivity and control can be achieved through cloud platforms, local gateways, and other methods. At the same time, data security and privacy protection need to be considered to ensure that user data is not leaked or abused.

(3) Function Design

In terms of function design, various smart home functions need to be designed according to actual needs. For example, smart security, smart lighting, smart home appliance control, voice control, etc. At the same time, the practicality and ease of use of the functions need to be considered so that users can easily use various functions.

(4) User Experience

In terms of user experience, attention should be paid to interface design, operation flow, response speed, and other aspects. A simple and clear interface design can be adopted to provide an intuitive operation flow, ensuring that users can quickly get started and enjoy the convenience brought by smart homes. At the same time, the compatibility and stability of devices need to be considered to ensure that users do not encounter various problems during use.

(5) After-sales service

In terms of after-sales service, it is necessary to provide comprehensive after-sales service and technical support. For problems or failures encountered by users during use, timely solutions and technical support need to be provided. At the same time, issues such as device repair and replacement need to be considered to provide users with more comprehensive after-sales service.

Based on the customer requirements mentioned in the previous section, we can determine:

① The hardware required for the system includes: fingerprint door lock, brightness sensor, infrared sensor, temperature and humidity sensor, CO concentration detector, smoke concentration detector, alarm device, control panel. Smart switches need to be installed for indoor ceiling lights, curtains, and air conditioning.

② Each terminal connects to the control panel through Bluetooth communication, constantly detects, and periodically sends the collected data to the control panel. The control panel sends the collected data to the control panel, receives instructions from the control panel, and sends them to each terminal.

③ The console connects to the network through Wi-Fi or wired connection and sends/receives messages to/from the dedicated cloud platform.

④ The design interface of the control panel should be able to display the collected data from the terminal and buttons for sending commands.

⑤ Users use mobile terminals and a proprietary app to access information from the cloud platform, view the indoor situation, and send control commands.

⑥ When the indoor temperature and humidity reach the threshold, the air conditioning and ventilation systems will automatically turn on or off.

⑦ When abnormal situations occur indoors, such as abnormal door lock opening, CO or smoke concentration exceeding the threshold, the alarm system will be activated, and an alert message will be sent to the owner's mobile device for notification.

The system architecture diagram is as Figure 3.1.2.

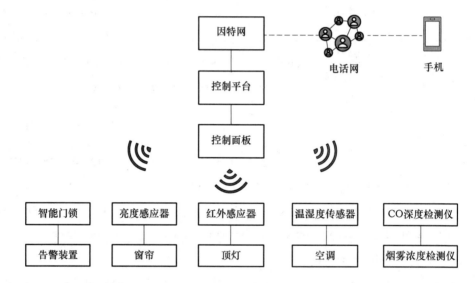

Figure 3.1.2　Architecture Diagram of Smart Home System

3. Question and Thinking

Question 1：What type of wireless access is more suitable for smart home scenarios? Why?

Question 2：What problems might be encountered when deploying a smart home system? How can they be resolved?

3.1.2　Smart Community

In northern cities in our country, there are many problems with winter heating, mainly manifested in uneven heating in residential households, high fuel consumption, and low thermal energy efficiency, and simple maintenance but high operating costs of the heating system. The country and the government are constantly exploring better models to solve these problems. According to the latest revised draft of the "Urban Heating Service (Solicitation of Comments)" issued in 2023, it is stipulated that under normal weather conditions and normal operation of the heating system, heating service companies should ensure that the indoor heating temperature of heating users meets the standards. Among them, the indoor heating temperature of residential and office buildings should not be lower than 18 ℃, the indoor heating temperature of nursing homes should not be lower than 22 ℃, and the indoor heating temperature of childcare institutions should not be lower than 20 ℃. At the same time, the draft also provides target values for evaluation indicators of heating service systems, requiring the timely response rate of heating equipment repairs to reach 100％, the timely handling rate of complaints to reach 100％, the timely handling rate of repair requests to be ≥98％, the qualified room temperature rate to be ≥98％, and the user satisfaction rate to be ≥96％.

With the development and maturity of information technology, smart communities play an important role in urban services and management. Smart communities fully utilize information technologies such as big data, cloud computing, and artificial intelligence to integrate various community service resources and create a new form of community governance based on informatization and intelligent management and services. Intelligent heating is an important scenario in smart community development.

1. Scenario Analysis

Due to the problems of high energy consumption, low efficiency, and high pollution in urban centralized heating, a large amount of energy resources is wasted, which seriously affects the environmental quality of cities. Energy saving and consumption reduction in urban centralized heating have become a key area for energy saving and emission reduction in China. In order to reduce energy consumption and achieve energy saving and emission reduction goals, heating companies are constantly exploring efficient and environmentally friendly heating methods, combining the heating industry with information technology, and using emerging technologies to replace traditional models, promoting the upgrading and transformation of the heating industry, and realizing the inevitable trend of intelligent heating development.

Intelligent heating mainly refers to the development based on information infrastructure, using technologies such as big data, artificial intelligence, cloud computing, and the

Internet. By collecting and analyzing heating-related data, intelligent control is applied to various heating processes, including heat source, heat network, and endpoint, in order to optimize the allocation of heat network resources and improve the accuracy of heat transmission. Obviously, the goal is to ensure the safe operation of heating equipment and achieve comprehensive heating control and operation through the use of big data, cloud computing, and automatic control, while maximizing energy conservation and environmental protection.

The importance of data collection and analysis feedback is evident in the process of intelligent heating, especially in the monitoring and collection of facilities and temperatures in residents' homes at the end of heating. Not only can it reflect the heating effect at the end, but the data can also serve as a strong basis for analyzing and adjusting heating parameters. See the Figure 3.1.3 below.

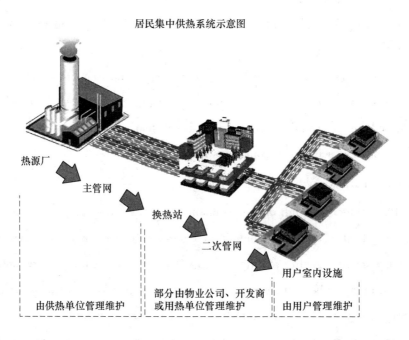

Figure 3.1.3 Schematic Diagram of Centralized Heating System

What other aspects and data need to be improved and monitored? To achieve the characteristics of a heating system that is monitorable, adjustable, measurable, and predictable, and to realize the green, safe, economical, and efficient operation of the system, intelligent heating should include aspects such as intelligent scheduling, intelligent regulation, intelligent control, intelligent diagnosis, intelligent maintenance, intelligent management, and intelligent services. They all have their own connotations.

① Intelligent Scheduling: The system can accurately forecast the heat load by automatically analyzing the thermal load characteristics of each building and then adjusting the dispatching plan of heat sources, heating networks, heat substations, and heat users according to the selected appropriate operating scheduling mode.

② Intelligent Control System for Heating Network: Smart Heat Network Operation and Maintenance System.

③ Intelligent Regulation: Heat users can control the room temperature according to their own needs.

④ Intelligent Diagnosis: By collecting and analyzing heating data, abnormal situations can be diagnosed in a timely manner, and relevant reminders can be given.

⑤ Intelligent Maintenance: Through the use of geographic information system technology and intelligent processing algorithms, intelligent scheduling, path planning, service quality, and assessment analysis can be applied to inspection, repair, maintenance, and other production maintenance activities.

In the above operational processes, it is necessary to comprehensively improve traditional inefficient work methods. The heating system mainly consists of three components: heat source, heat network, and end terminals. To ensure the normal operation of each component, relying solely on traditional management and operation methods not only results in low work efficiency but also leads to resource waste, which does not meet the current needs of sustainable development.

For example, in the operation of the heating pipeline network, it is necessary to conduct daily inspections and maintenance of the pipeline network and regulate the heat exchange stations between the networks. In the past, pipeline maintenance was done manually, which easily led to overlooked problems and potential hazards in the future. Additionally, if the heat exchange allocation in the heat exchange stations is uneven, it can result in excessively high or low heating temperatures in the community. This requires the collection, analysis, and processing of other relevant data, apart from the end terminals, to support efficient and precise automated control.

Taking the heating monitoring system in X City as an example, the system needs to meet the following requirements.

Requirement 1: Terminal 1 can collect indoor temperature data in real-time and report it to the platform; Terminal 2 can report the status of pipeline valves in real-time.

Requirement 2: The platform can display real-time data, historical data, and current management valve information reported by each terminal.

Requirement 3: The platform can set temperature thresholds for different areas.

Requirement 4: When the platform receives abnormal data, it can automatically send out alarm messages and independently balance and control the heating equipment in the abnormal area.

2. Project Scheme Design

The design of the intelligent heating monitoring system needs to consider the accuracy and stability of data collection, the reliability and security of remote control, the timeliness and accuracy of intelligent alarms, the usability and intuitiveness of graphical data interfaces, and the complexity and maintainability of system management.

According to system requirements, we need to consider the following factors and indicators during the design process.

① Determining the monitoring scope and parameters: Based on the scale and requirements of the heating system, determine the scope of monitoring, such as heating pipelines, thermal nodes, etc., as well as the parameters to be collected, such as temperature, pressure, and flow rate, etc.

② Selecting suitable sensors and devices: Based on the requirements of the monitoring scope and parameters, select suitable sensors and devices, such as temperature sensors, pressure sensors, flow meters, etc., as well as data collectors and network devices.

③ Designing data transmission and storage solutions: Determine the method and frequency of data transmission, select suitable data storage devices and technologies to ensure the real-time and security of the data.

④ Developing Monitoring System Software: According to the requirements of the heating monitoring system, develop corresponding software systems, including data collection, processing, analysis, display, and alarm functions.

⑤ Integration and Debugging of the System: Integrate and debug the selected equipment and software to ensure the stability and reliability of the system.

⑥ Deployment and Operation of the System: Deploy the heating monitoring system to the actual operating environment for daily operation and maintenance, ensuring the normal operation of the system and the accuracy of the data.

The system architecture is shown in the Figure 3.1.4 below.

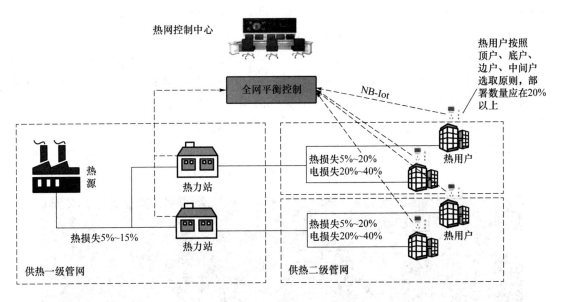

Figure 3.1.4 Architecture Diagram of Centralized Heating System

Due to the wide coverage, small data volume, low latency requirements, and certain requirements for the privacy and security of some data, the heating monitoring system adopts NB-IoT mode technology for wireless access.

The collection terminal uses the Qingping wall-mounted thermometer, as shown in Fiture 3.1.5, which can display the temperature and humidity of the test environment in real-time. It supports NB-IoT and can be powered by mains electricity. It has a dedicated APP for remote setting of the sampling frequency. The device itself can store 2 880 sets of data, so there is no need to worry about data loss when the network is disconnected.

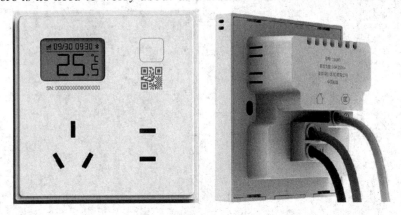

Figure 3.1.5 Qingping Wall-mounted Thermometer

The platform uses the exclusive cloud platform provided by the operator to store the data reported by the terminals connected to the network in the area. It allows for the viewing of temperature information for each household in different cities, communities, and units. It can monitor the opening and closing status, flow, and temperature of the heating pipes, and remotely control the switches. It also provides alarm reporting and reminders in case of data anomalies.

The client of the heating network control center is a dedicated webpage developed by the heating company for data viewing and device management, as shown in Figure 3.1.6 ～ Figure 3.1.9. It can view various data reported by indoor residents and heating network valves, set threshold parameters for different areas, and provide alerts for abnormal data.

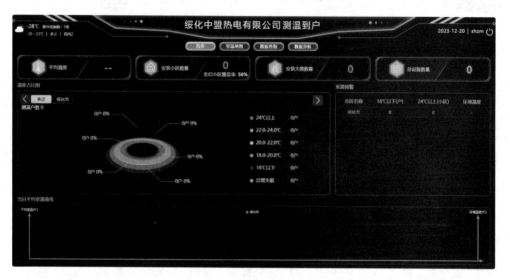

Figure 3.1.6　Home Page of the Heating Network Control Center Client

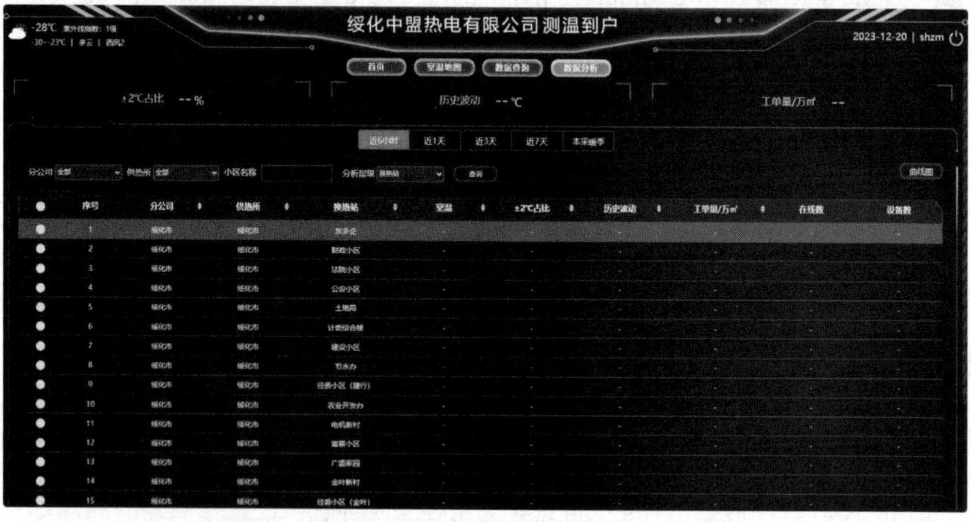

Figure 3.1.7　Community View Page of the Heating Network Control Center Client

Figure 3.1.8 Resident View Page of the Heating Network Control Center Client

Figure 3.1.9 Residential Detailed Information Page of the Heating Network Control Center Client

3. Implementation Plan

The basic implementation process includes:

(1) Equipment installation and debugging

First, the installation and debugging of thermometers and smart switches are required. The installation process should be carried out according to the equipment manual to ensure correct installation and connection of the devices. At the same time, the devices need to be debugged to ensure normal operation and functionality.

(2) Network activation and debugging

After the installation of the devices, SIM card registration and installation need to be carried out for the terminals. The NB-IoT related configurations should be done in the background of the base station, and tests should be conducted to confirm network coverage and signal indicators. This includes device connection, control, and data transmission. The system needs to be debugged to ensure smooth communication and collaboration among the devices. Additionally, system optimization is required to improve stability and efficiency.

(3) User Training and Usage Guidance

After the installation of the equipment and system integration is completed, it is necessary to provide training and guidance to end users, including residents using the equipment and maintenance personnel from the heating company using the monitoring system. The guidance should include the methods of using the equipment, introduction to its functions, operating procedures, and other aspects.

4. Solution Validation

The validation of the heating monitoring system design can be done through the following steps.

① Requirement validation: By communicating with users, conducting surveys, and using other methods, the accuracy and completeness of the requirement analysis are verified to ensure that the heating monitoring system can meet the users' needs.

② Sensor validation: In the actual environment, selected sensors are tested to compare their measurement results with the actual values in order to validate the accuracy and reliability of the sensors.

③ Data Collection and Transmission Verification: By examining the frequency, size, and integrity of data collection and transmission, the accuracy and reliability of data collection and transmission can be verified.

④ Data Processing and Analysis Verification: By comparing the differences between processed data and original data, or by comparing the data trends during different time periods, the accuracy and reliability of data processing and analysis can be verified.

⑤ User Interface Verification: Inviting real users to use the user interface of the heating monitoring system in actual environments, collecting user feedback and opinions to verify the usability and visual effects of the user interface.

⑥ System Integration and Debugging Verification: By conducting various tests and verifications, such as functional testing, performance testing, and security testing, the accuracy and reliability of the integration and debugging of the heating monitoring system can be verified.

⑦ System Deployment and Operation Verification: In the actual operating environment, the heating monitoring system is run and monitored for a long period of time, collecting operational data and user feedback to verify the stability and reliability of the system.

Through the above steps, a comprehensive validation of the design of the heating monitoring system can be conducted to ensure its accuracy and reliability, improve heating efficiency, reduce energy waste, and enhance user satisfaction.

Here, we will not elaborate on the installation and debugging process of the terminal. The experiment focuses on the configuration and opening process of the NB-IoT base station. The base stations used are ZTE's B8200 and R8872A, which are commonly used by operators. There are four types of configurations for base station data: global data, equipment data, transmission data, and wireless data.

Experiment 1   Configuring Global Data

Including creating subnets and network elements, creating operators and PLMN, as shown in Figure 3.1.10~Figure 3.1.11.

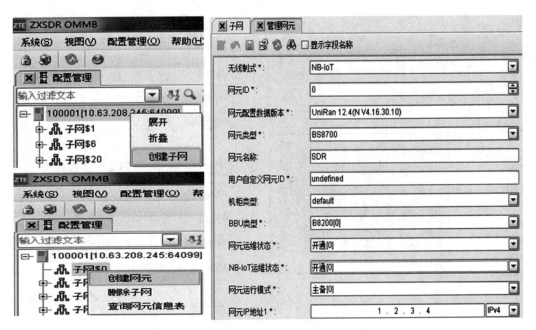

Figure 3.1.10   Manage Network Element Configuration Page

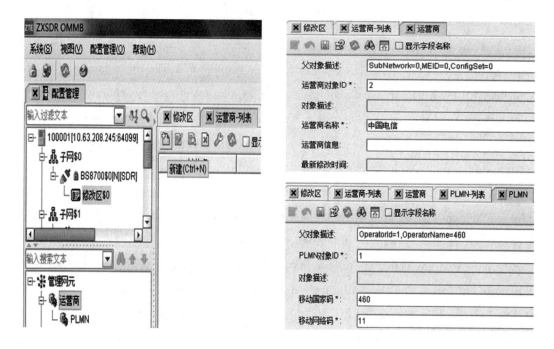

Figure 3.1.11 PLMN Configuration Page

Experiment 2 Configuring Device Data

Including creating BBU hardware boards, creating RRU hardware boards, and creating RRU topology, as shown in Figure 3.1.12~Figure 3.1.15.

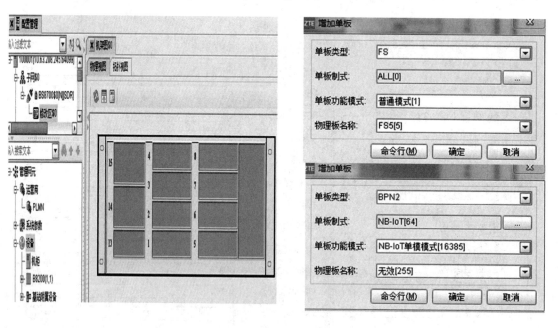

Figure 3.1.12 Single Board Configuration Page

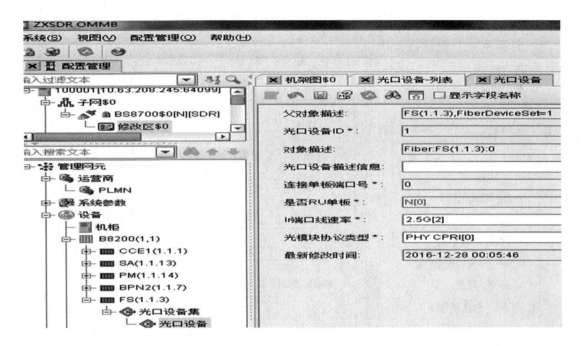

Figure 3.1.13 Optical Port Device Configuration Page

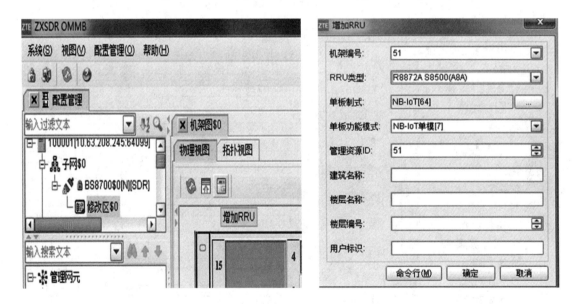

Figure 3.1.14 RRU Configuration Page

Experiment 3　Configuring Transmission Data

Including creating physical layer ports, creating Ethernet links, creating IP configurations, creating bandwidth resources, creating SCTP, creating business and DSCP mappings, and creating OMC channels, as shown in Figure 3.1.16～Figure 3.1.22.

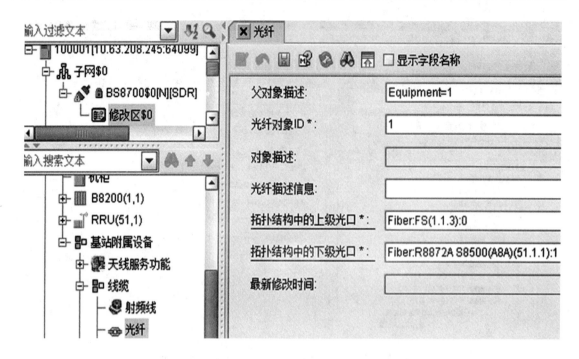

Figure 3.1.15 Fiber Optic Configuration Page

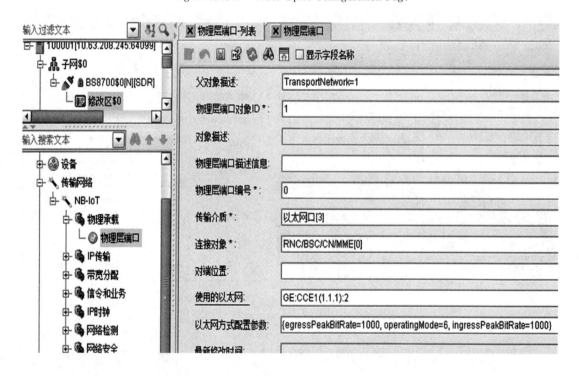

Figure 3.1.16 Physical Layer Port Configuration Page

Figure 3.1.17　Ethernet Link Layer Configuration Page

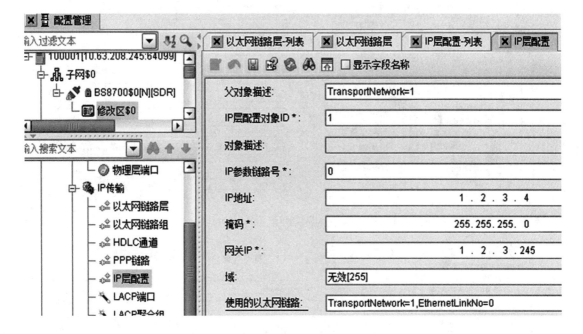

Figure 3.1.18　IP Layer Port Configuration Page

Figure 3.1.19　Bandwidth Configuration Page

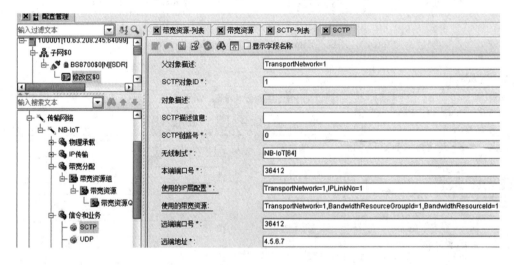

Figure 3.1.20　SCTP Configuration Page

Figure 3.1.21　Business and DSCP Mapping Configuration Page

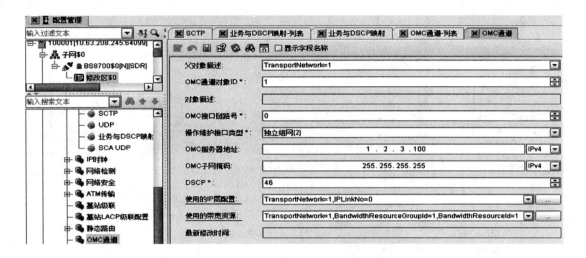

Figure 3.1.22　OMC Channels Configuration Page

Experiment 4　Configuring Wireless Data

Including creating N-BIOT wireless nodes, creating N-BIOT baseband resources, creating N-BIOT basic cell information, and creating N-BIOT cell reselection relationships, as shown in Figure 3.1.23～Figure 3.1.26.

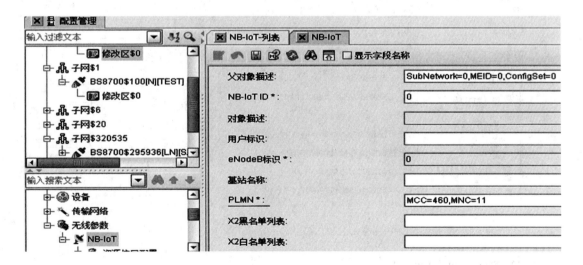

Figure 3.1.23　Site Configuration Page

• 89 •

Figure 3.1.24 Baseband Configuration Page

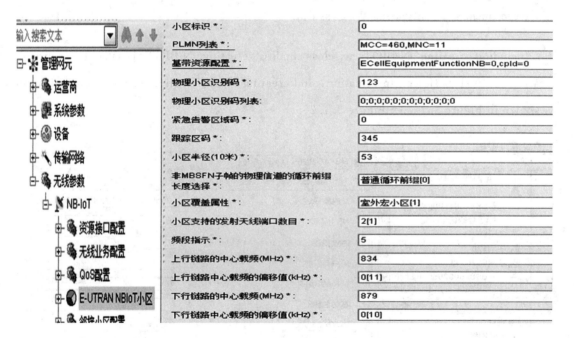

Figure 3.1.25 Cell Configuration Page

5. Question and Thinking

Question 1:Besides heating systems, what other scenarios in a smart community could use NB-IoT wireless deployment?

Question 2:Apart from NB-IoT, what other wireless access methods are suitable for the scenarios discussed in this chapter?

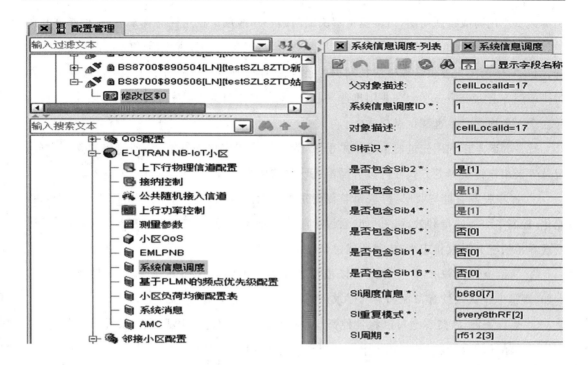

Figure 3.1.26　System Information Scheduling Configuration Page

3.1.3　智慧养老

随着我国现代经济社会的高速发展,科学生产技术的进步,我国人民的生活质量相比过去正变得越来越好,卫生水平、医疗条件都得到了很大的改善和提高,我国老年人口的平均寿命明显延长,人口的年龄结构逐渐呈现出人口老龄化的发展特征。我国在 1999 年进入老龄社会,老年群体增长速度快、数量多。《2018 年民政事业发展统计公报》指出,截至 2018 年底,我国年满 60 周岁以及超过 60 周岁的老年人为 24 949 万人,占总人口的 17.9%。数量庞大的老年群体和老年人口较快的增长速度是现阶段我国人口老龄化十分重要的特征。由于我国老龄化具有人口基数大、增速快、高龄化三大特征,加之我国正处于社会主义初级阶段的基本国情,在经济上仍然处于"三期叠加",且城市化进程加快发展导致的诸多社会问题未能解决,如何解决我国养老问题已经成为当前政府改善民生、提升人民幸福感的首要任务。

我国目前的养老形式大体可以概括为"9073",代表了我国现行的三种养老模式,"90"代表 90% 身体健康状况较好的老年人选择家庭养老;"7"代表 7% 的老年人选择社区养老;"3"代表 3% 生活不能自理的老年人选择养老机构。随着我国人民生活水平提高,老年人对养老服务的需求也发生了很大的变化。家庭养老不能满足老年人的多种养老需求;社区养老的养老服务设置不合理、设施简陋、服务水平低、资源利用不充分;公办养老服务机构一床难求,私立养老机构或费用过高或设施简陋,诸多亟待解决的问题一一显现。在这样的背景下,一种能够利用计算机网络技术、物联网信息技术等现代科学技术,与老年人在日常生活中的各个方面紧密结合,为每一位老年人提供诸如生活长期护理照料、健康管理照护等多种养老服务的移动互联网时代智慧养老综合服务平台在我国迅速兴起,该平台主要服务于居家老年人和社区养老机构,并在全国各地逐步推进和试点探索。

智慧养老是指利用先进的信息技术手段,为老年人提供更加智能化、便捷化的养老服务。智慧养老相比传统养老方式在许多方面都有明显的优势:

① 提高服务质量与效率:智慧养老利用先进的信息技术,实现了服务流程的自动化和智能化,提高了服务质量和效率。例如,通过智能化设备实时监测老年人的健康状况,可以及时发现异常情况并采取相应措施,避免错过最佳治疗时机。

② 满足个性化需求:智慧养老可以根据老年人的不同需求和偏好,提供个性化的服务。例如,根据老年人的饮食习惯和营养需求,为其推荐合适的菜谱和食品。

③ 整合资源:智慧养老可以将各种资源进行整合,包括医疗、生活照料、文化娱乐等,方便老年人一站式获取所需服务,减少中间环节和信息不对称的问题。

④ 降低成本:智慧养老利用智能化设备和数据分析,可以减少人力成本和资源浪费,从而降低养老服务的成本。另外,通过提高服务效率和质量也可以降低医疗成本和其他相关成本。

⑤ 增强互动与沟通:智慧养老为老年人提供了更加便捷的沟通方式,如视频通话、在线交流等,方便老年人与家人、朋友、服务机构之间的交流和互动。

⑥ 促进创新:智慧养老的发展推动了技术创新和服务创新,为养老产业的发展注入了新的活力。例如,智能家居、智能健康监测等设备的研发和应用,推动了智慧养老技术的不断进步。

当然,智慧养老也存在一些问题。一方面,智慧养老服务体系不断完善。政府、企业和社会组织都在积极推动智慧养老的发展,通过建设智慧养老服务平台,推广智能化养老设备等方式,为老年人提供更加便捷、高效的服务。例如,智能家居、智能健康监测等设备可以帮助老年人更好地管理健康,提高生活质量。另一方面,智慧养老服务市场还存在一些问题。首先,老年人的信息素养普遍较低,对智慧养老服务的接受程度有限。其次,智慧养老服务的质量和安全性需要进一步提高。由于智慧养老服务涉及众多领域,如医疗、生活照料等,不同领域之间的数据共享和整合存在一定难度,导致服务质量和效率受到影响。此外,智慧养老服务也存在一定的安全风险,如个人信息泄露、设备故障等。

针对这些问题,政府和企业需要加强合作,加大对智慧养老服务的投入和监管力度。同时,也需要加强宣传教育,提高老年人的信息素养和对智慧养老服务的认知度。未来,随着技术的不断进步和市场的不断扩大,相信智慧养老服务会得到更加广泛的应用。

综上所述,智慧养老具有服务优质、高效便捷、个性化定制等优势。通过应用智慧养老,可以提高老年人的生活质量,促进社会和谐发展。

1. 场景分析

智慧养老的场景主要包括以下 6 个方面:

① 健康管理:通过智能设备监测老年人的健康状况,如心率、血压、血糖等,并将数据上传至云端进行分析,为老年人提供个性化的健康管理方案。

② 居家照护:利用智能家居设备,如智能音箱、智能照明等,为老年人提供便捷的生活照护服务,如定时提醒吃药等。

③ 社区服务:通过智慧养老服务平台,为老年人提供社区服务,如家政服务、健康讲座、文化娱乐等。

④ 紧急救援:通过智能设备监测老年人的生命体征,一旦监测出现异常情况,可以及时发出警报并联系医疗机构进行紧急救援。

⑤ 智能支付:智慧养老服务平台为老年人提供线上支付功能,方便老年人进行各类费用的缴纳。

⑥ 远程关怀:智慧养老服务平台为异地子女提供远程关怀服务,如视频通话、查看父母健康数据等。

以上是智慧养老的一些常见场景,随着技术的进步,智慧养老场景将更加丰富和多样化。

爱窝云提出一种主动式居家养老服务模式,既能保证老年人在熟悉的家庭环境中生活,又能保证在遇到特殊情况时能够及时得到就医。该模式服务的对象是老年人(失能、半失能、高龄、空巢老人),处于术后恢复期患者、需要长期疗养的慢性病人、行动不便的残障人士,服务模式的目标是此类人群通过手机操作,就可以在家获得医疗级的服务,包括居家护理、医学康复、远程诊疗、慢病管理、康复护理、专项护理、健康教育等。使用手机实现养老服务全流程如图 3.1.27 所示。

图 3.1.27 手机实现养老服务全流程

以居家健康管理为例,服务对象是患有慢性病的居家老年人,需要定时、定期进行个人健康数据监测、远程问诊、健康随访、慢病管理、复诊提醒及健康宣教。

① 采集终端。需要健康监测设备,包括智能手环、智能手表、智能体重秤等,可以监测人体的生理参数,如心率、血压、血糖、血氧等,并将数据传输到健康管理平台。

② 云平台。可以查看服务对象的实时和历史健康数据,根据数据提供相应的健康建议,并在数据异常时进行风险预警和提醒。

③ 客户端。面向服务对象的 APP 和面向医务人员的 APP。面对服务对象的 APP 上可以记录健康状况,包括日常饮食、运动情况、睡眠情况等,根据个人情况显示个性化的健康建议,并在需要医疗服务时可以进行服务下单和一键呼救;面向医务人员的 APP 上可以进行数据查看、视频问诊等。

2. 项目方案设计

居家健康管理的方案设计需要考虑以下 7 个方面:

① 确定目标:明确健康管理的目标,如降低患病风险、提高生活质量、控制慢性病等。

② 评估现状:了解个人的健康状况和需求,包括生理参数、生活习惯、家族病史等。

③ 制定计划:根据个人情况制订个性化的健康管理计划,包括饮食、运动、心理等方面的建议。

④ 选择采集终端:根据个人需求选择适合自己的采集终端,如智能手环、智能手表、健康管理手机 APP 等。

⑤ 监测数据:通过采集终端监测个人的生理参数和健康状况,并及时上传数据到健康管理平台。

⑥ 分析评估:健康管理平台对上传的数据进行分析和评估,并提供个性化的健康建议和干预措施。

⑦ 调整计划:根据个人情况和健康建议调整健康管理计划,并持续监测和评估。

根据场景分析,爱窝云提出养老服务方案架构如图 3.1.28 所示。

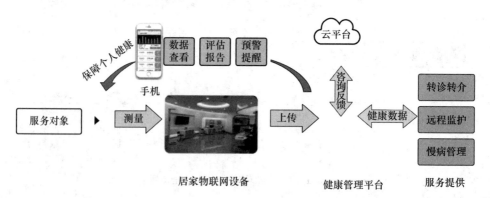

图 3.1.28 养老服务方案架构

本方案中的采集终端使用智能手环、血压计、睡眠仪、生化仪,实时监测服务对象的基础生理参数,如心率、血压、血糖、睡眠情况等;智能家居设备使用智能温度计和智能空气净化器,以监测室内环境温湿度、空气质量。

云平台使用爱窝云自研的平台,进行慢病病程监控及并发症风险预警。平台可以存储服务对象的健康档案、基础生理数据、重点监测指数、异常数据信息,并通过平台自身的云计算功能进行健康评级,提出医疗建议和风险预警,如图 3.1.29 和图 3.1.30 所示。

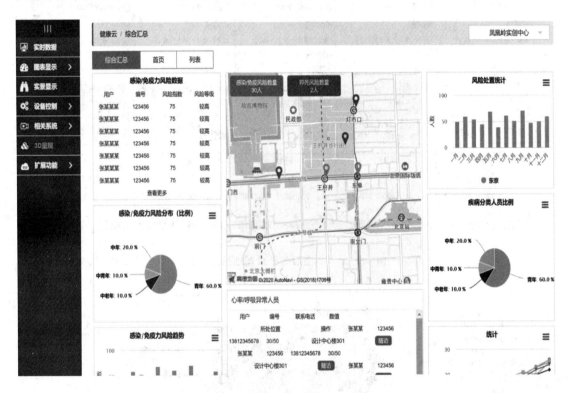

图 3.1.29　养老服务平台主页

客户端使用爱窝云自研的 APP,能够显示服务对象的基础生理数据和健康状况,提供个性化的健康建议,以及异常指标风险提醒,如图 3.1.31 所示。

3. 问题与思考

问题 1:在智慧养老场景中,一些安全隐患,如诈骗、滑倒、火灾等问题,威胁着老年人的安全,应如何预防这些潜在的安全风险?

问题 2:在智慧养老场景中,老年人的身体健康数据需要开放给健康管理机构,应如何确保这些数据的安全性和保密性?

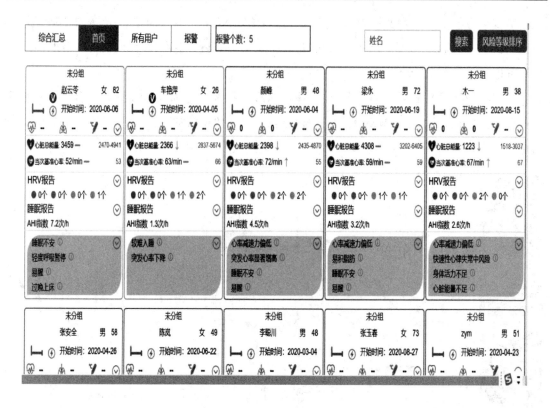

图 3.1.30 养老服务平台详细信息页面

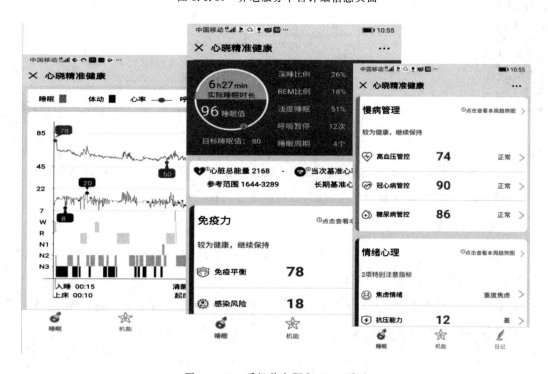

图 3.1.31 手机养老服务 APP 页面

3.2 绿色双碳

3.2.1 算力中心环境监测

算力中心是一种集中计算资源的服务平台,它为用户提供高性能的计算资源和数据处理服务。随着数字化时代的到来,算力中心在各个领域的应用越来越广泛。

受政策与市场需求双重驱动,算力中心目前在我国呈现高速增长态势,规模不断扩大。截至 2023 年 6 月底,全国在用数据中心机架总规模超过 760 万标准机架,算力总规模达到 197EFLOPS,位居世界第二。

随着大模型、AIGC 等人工智能应用的快速发展,社会总体算力需求中的智算需求进一步扩大,并成为引领算力中心发展的重要驱动。北京、上海、深圳等城市都在积极推动智算中心建设。

此外,我国还在积极推动算力融合应用,加速涌现出新的应用场景。同时,为了满足算力需求,数据中心的"降耗增效"也在紧锣密鼓地进行中。政府和企业都在积极探索优化数据中心制冷系统,提高算力应用环节效率。

然而,需要注意到的是,大模型产业井喷式发展的同时带来了算力紧缺、能耗激增等问题。目前国内已有 100 多个大模型公开发布,加剧了智能算力紧张的局面。

为了确保算力中心高效、稳定运行,提高整个系统的可靠性、稳定性和安全性,需要对算力中心环境进行环境监测,监测系统的主要作用在于:

① 保障设备运行:算力中心的设备对环境条件有一定要求,如温度、湿度、洁净度等。通过环境监测,可以实时了解这些参数,确保设备在适宜的环境中运行,延长设备使用寿命,减少故障率。

② 提高能效:通过监测环境温度、湿度等参数,可以合理调节算力中心的空调系统和其他能耗设备,降低能耗,提高能效,以减少能源浪费和运营成本。

③ 预防安全隐患:一些环境因素(如湿度过高或过低)可能导致设备故障、电路短路等问题,从而引发安全隐患。进行环境监测有助于及时发现异常,预防安全事故的发生。

④ 优化运维管理:通过对环境的实时监测和数据记录,可以全面了解算力中心的环境状

况,为运维人员提供参考以及制订更合理的维护计划,提高运维效率。

⑤ 促进技术创新:环境监测是算力中心技术体系的重要组成部分。通过持续的环境监测实践和技术研究,可以促进相关技术的创新和发展,推动算力中心整体技术水平的提升。

⑥ 辅助决策支持:实时、准确的环境监测数据可以为管理层提供决策依据。例如,可以根据环境变化趋势预测设备维护需求,调整能源消耗策略等。

⑦ 满足法规要求:在一些国家和地区,对数据中心的环境有明确的法规要求。环境监测有助于确保算力中心符合相关法规标准,降低合规风险。

⑧ 提升客户满意度:对于提供算力服务的第三方数据中心而言,为客户提供稳定、高效的服务是至关重要的。通过环境监测来确保优质的服务体验,有助于提高客户满意度,赢得市场信任。

综上,监测算力中心环境不仅有助于确保设备的正常运行和降低能耗成本,还可以提升系统的可靠性和安全性,促进运维管理的优化和技术创新。对于数据中心的管理者和运营者来说,重视并实施有效的环境监测策略具有重要意义。

数据中心环境监控系统在实际应用中有很多成功案例,以下是一些典型案例:

① 腾讯数据中心环境监控系统:腾讯数据中心采用了先进的环境监控系统,可以实时监测数据中心的温度、湿度、气压、气体成分等环境参数,以及设备的运行状态和能耗情况。该系统能够及时发现异常情况并发出警报,同时自动调节环境参数,确保设备的正常运行和数据的可靠性。

② 阿里巴巴数据中心环境监控系统:阿里巴巴的数据中心环境监控系统采用了物联网技术和大数据分析技术,对数据中心的温度、湿度、洁净度、气体成分等环境参数进行实时监测和分析。该系统能够预测环境参数的变化趋势,并提供优化建议和预警信息,帮助管理者制定节能策略和应对措施。

③ 百度数据中心环境监控系统:百度数据中心的环境监控系统采用了智能传感器和自动化控制技术,对数据中心的温度、湿度、气压、气体成分等环境参数进行实时监测和调节。该系统能够根据实际需求自动调节环境参数,提高数据中心的能效和可靠性,降低能耗和维护成本。

除了上述案例,还有很多知名企业(如亚马逊、谷歌、微软等)也都在数据中心环境监控方面投入了大量资源和应用。这些企业的成功应用案例都表明,数据中心环境监控系统在保障设备正常运行和数据可靠性方面发挥着重要作用,同时也有助于降低能耗和维护成本,提高数据中心的运行效率和管理水平。

1. 场景分析

要实现对算力中心的环境监测,需要实现以下功能:

① 实时监测:系统可以实时监测算力中心内的温度、湿度、气压、洁净度、气体成分、水浸环境参数,工作设备、空调设备、照明设备、其他用电和电源系统的运行状态和能耗情况,以及算力中心的热量分布。

② 报警功能:当环境参数超过预设阈值或设备能耗出现异常时,系统能够及时发出警报并通知相关人员处理,避免设备损坏和数据丢失。

③ 控制功能:系统可以根据预设的策略自动调节环境参数,如自动调节空调系统的温度、湿度等,以确保设备运行在最佳状态。

④ 能源管理:系统可以对算力中心的能源使用情况进行监测和管理,提供能源消耗的统计和分析数据,帮助管理者制定节能策略和优化方案。

⑤ 安全管理:系统可以对数据中心的安全状况进行监测和管理,如视频监控、门禁控制等,以确保数据中心的安全运行。

要实现以上功能,需要具备的采集终端包括:温湿度传感器、水浸采集器、气体检测传感仪、电量仪。

算力中心一般面积较大,如OPPO全球算力中心占地面积约为106.5亩,中国电信中南智能算力中心项目规划占地面积约300亩,长春算力中心项目占地面积5 280 m^2,所以采集终端设备的数量庞大、能耗较高且对通信距离要求较高,宜用NBIOT或LoRaWAN方式进行无线接入云平台。

云平台需要具备以下功能:

① 能够分区域查看各采集器的实时上报数据、历史数据,并生成各数据趋势图;

② 能够分区域设定各类数据的阈值,当数据异常或超出阈值时能够进行告警提醒;

③ 能够分区域获取各能耗系统的电源开关状态,并远程控制开关设备;

④ 能够根据能耗趋势给出相应的节能建议,并发送至客户端。

云平台发送数据至专属客户端,算力中心管理人员可以对客户端进行登录、查看等操作。

2. 项目方案设计

算力中心环境监控方案设计需要综合考虑算力中心的特点和需求,包括数据中心的规模、布局、设备配置、能源效率等方面。根据场景分析,一个算力中心监测系统的架构如图3.2.1所示。

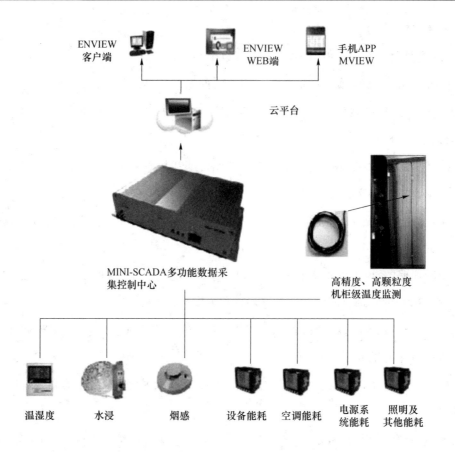

图 3.2.1 算力中心监测系统架构图

方案中用到的终端包括：

① MINI-SCADA 多功能数据采集控制中心：使用市电供电，可挂接多个传感器、电量仪，对下挂的设备进行统一的数据上报。

② 三维热量数据采集单元：采集现场测量点的温度值，以"一线总线"的数字方式传输，大大减少系统的电缆数量。

③ 壁挂 LCD 显示温湿度传感器：LCD 可以实时显示采集的测量环境的温湿度。

④ 电量仪：实时采集测量点的电量使用情况。

⑤ 光电烟感控测器：对开放式区域提供烟雾环境保护，环境异常时发出告警提醒。

⑥ 水浸传感器：对测量点进行水浸检测，异常时发出告警提醒。

云平台和客户端能够实时显示算力中心的各项数据，如图 3.2.2～图 3.2.4 所示，分别是环境数据图、能耗数据图、三维热量图。

图 3.2.2　算力中心监测系统环境数据图

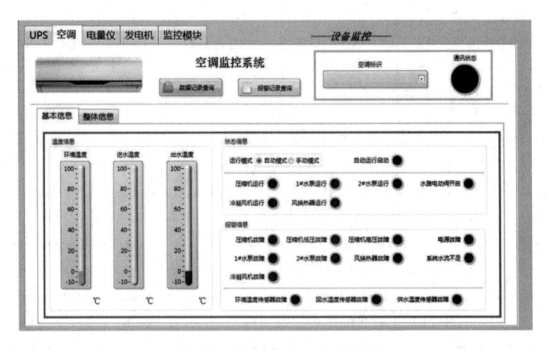

图 3.2.3　算力中心监测系统能耗数据图

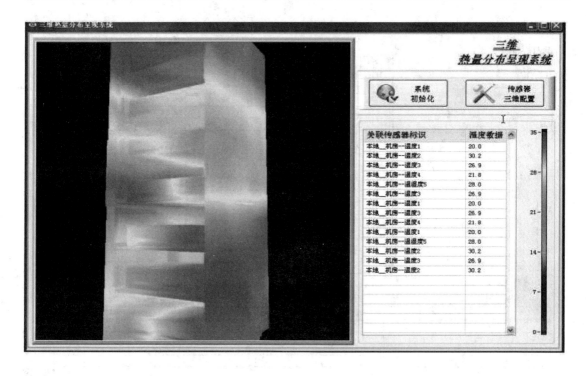

图 3.2.4 算力中心监测系统三维热量图

3. 问题与思考

问题：算力中心的设备是 7×24 h 不停运行的，对算力中心的环境监控也是如此。那么，如何在保证监控系统正常工作的前提下降低系统能耗，以满足算力中心绿色环保的要求呢？

3.2.2 Building Safety

Concrete, as a commonly used material in construction projects, has a significant impact on the safety and stability of buildings in terms of its performance and quality.

Heat of hydration is a common issue in concrete construction, and the temperature change caused by the heat released during the hydration process of concrete is referred to as the heat of hydration temperature. Concrete hydration refers to the chemical reaction between water and cement, resulting in the formation of hydration products. During this process, a large amount of heat is released, leading to an increase in concrete temperature.

The heat of hydration temperature of concrete has a significant impact on its performance and service life. If the hydration temperature of concrete is too high, it can cause cracks and changes in the concrete, and even damage its structure. On the other hand, if the heat of hydration temperature is too low, it can affect the early strength development and frost resistance of the concrete.

The peak temperature of heat of hydration in concrete is usually between 40℃ and 70℃, but the specific temperature is influenced by various factors, including the mix ratio of concrete, hydration reaction rate, and environmental temperature. Generally speaking, the higher the water-cement ratio in the mix ratio of concrete, the higher the temperature of heat of hydration; the faster the hydration reaction rate, the higher the temperature of heat of hydration; and the higher the environmental temperature, the higher the temperature of heat of hydration.

Controlling the temperature of heat of hydration in concrete is also very important for construction. By monitoring the temperature of heat of hydration in concrete in real-time, measures can be taken in a timely manner to control the temperature changes in concrete, avoid cracks and deformations, and ensure the quality and safety of the concrete. This is mainly for the following aspects.

① Preventing Concrete Cracking: The heat of hydration can cause the internal temperature of concrete to rise. If not properly controlled, this can result in temperature stress and lead to concrete cracking. By monitoring the temperature due to heat of hydration, measures can be taken in a timely manner to reduce the temperature and prevent cracking.

② Controlling Concrete Quality: Both excessively high and low temperatures due to heat of hydration can impact on concrete quality. Excessively high temperatures can lead to issues such as cracking and deformation, while excessively low temperatures can affect the strength and durability of concrete. Therefore, monitoring the temperature due to the heat of hydration is helpful in controlling concrete quality.

③ Ensuring Construction Safety: Excessive heat of hydration can lead to cracks in concrete, which can affect structural safety. By monitoring the temperature due to heat of hydration, problems can be detected and addressed in a timely manner to ensure construction safety.

In terms of building safety, the heat of hydration is an important consideration, especially when it comes to monitoring the temperature of concrete during hydration.

1. Scenario Analysis

The Concrete Heat of Hydration Monitoring Platform is a system specifically designed to monitor and evaluate the heat generated during the hydration process of concrete. This platform is suitable for various large-scale infrastructure construction projects, such as dams, bridges, and high-rise buildings, with the aim of ensuring the integrity and safety of concrete structures. By deploying a network of temperature sensors inside the concrete structure, real-time temperature data can be collected. After processing and analysis, this data can provide engineers with key information about the hydration process, temperature changes, and potential risks of thermal cracking in the concrete.

The concrete hydration heat monitoring platform needs to achieve the following basic functions.

① Real-time monitoring: The platform needs to be able to collect, transmit, and process temperature data in real-time, so that engineers can timely understand the hydration heat situation of the concrete.

② Data visualization: The platform should provide intuitive data visualization tools to help engineers better understand and analyze temperature data.

③ Warning system: When abnormal temperature fluctuations or potential risks are detected, the platform should be able to send warning messages to engineers in a timely manner.

④ Historical data recording: The platform needs to save historical temperature data so that engineers can conduct long-term analysis and evaluation.

⑤ User Permission Management: The platform should have a comprehensive user permission management system to ensure the security and confidentiality of data.

Technical Implementation of Monitoring Platform for Heat of Hydration in Concrete.

① Sensor Network: Construct a network using high-precision temperature sensors to achieve real-time monitoring of the internal temperature of the concrete.

② Data Transmission: Use wireless communication technologies such as LoRa and NB-IoT to transmit temperature data from the sensor network to the data center.

③ Data Processing and Analysis: Use big data processing techniques and machine learning algorithms to process and analyze temperature data in real-time, extract key information, and generate reports.

④ Data Visualization: Develop a data visualization interface using web front-end technologies such as React and Vue to facilitate engineers in viewing and analyzing temperature data.

⑤ Early Warning System: Based on preset thresholds and rules, it constructs an early warning model to detect and notify of abnormal situations in a timely manner.

⑥ Database Management: It uses relational databases (such as MySQL, PostgreSQL, etc.) to store historical temperature data and user information, ensuring data persistence and security.

⑦ User Authorization Management: It adopts methods such as Role-Based Access Control (RBAC) to achieve flexible management and control of user permissions.

The Concrete Hydration Heat Monitoring Platform provides engineers with critical information about the safety and stability of concrete structures through real-time monitoring and analysis of internal temperature data. This platform has a wide range of applications and can play an important role in various large-scale infrastructure construction projects. With the continuous advancement of technology and changes in market demand, this platform still has great potential for development and innovation in the future.

2. Project Scheme Design

IoT Wireless High-Precision Concrete Hydration Heat Cloud Monitoring Platform, using high-precision (± 0.1 ℃) intelligent bus-type temperature sensors, combined with the on-site conditions in China, can choose from a variety of wideband and narrowband wireless communication technologies available on the market, such as 4G, NB-IoT, CAT1, LoRaWAN, etc. Due to the fact that the construction site is a special industrial park with a large area and the monitoring data has privacy, we have chosen to build a private LoRaWAN network for system monitoring.

The entire system only uses digital signal transmission, and the sensors and acquisition modules can be connected to the network, making the system more reliable and the wiring more convenient. The main functions of this system are:

① The wiring is minimal, the installation workload is small and convenient, and monitoring personnel can view the temperature changes of all monitoring points anytime and anywhere through a web browser or mobile APP, as long as there is network coverage in the area.

② By using high-precision digital temperature probes, the measurement accuracy is better than ± 0.2 ℃ (actual test data is better than ± 0.1 ℃), and the resolution is up to 13 digits, which is 0.031 25 ℃.

③ Continuous monitoring for 24 hours. When the temperature exceeds the set limit, the relevant parties, such as the client, supervisor, and construction team, will be promptly notified through different methods, such as mobile APPs, text messages, or emails.

④ The equipment is safe and reliable. The wireless acquisition module can be powered by a long-lasting battery (nominal life of more than 3 years), effectively preventing electrical leakage and electric shock accidents.

⑤ By dividing the measurement area into several parts through on-site acquisition modules, distributed measurement can be achieved. The communication between gateways adopts wireless (4G/LoRaWAN/CAT1/NBIoT) communication methods, effectively utilizing the excellent network expansion capability of wireless (4G/LoRaWAN/CAT1/NBIoT) networks. As the scale of the enterprise expands, there is no need to change the original system structure. It can be fully utilized and only requires targeted expansion to complete system capacity expansion.

⑥ The monitoring data is directly transmitted to the cloud through a wireless gateway, eliminating the need for a host on-site. This natural isolation ensures the safe operation of the monitoring room equipment and the personal safety of the staff on duty, effectively compared to wired solutions.

⑦ After the system collects data from various monitoring points to the backend database system in the cloud, it provides analysis and prediction functions through analysis and processing. At the same time, it provides multi-dimensional and visually convenient services to various stakeholders through web pages, mobile APPs, and other means.

⑧ This system has good scalability. In addition to temperature data, it can also monitor smoke alarms, ground ponding, water level alarms, theft and intrusion detection, or add other project monitoring according to customer needs.

⑨ Permission Division: The access permissions for various on-site data are set uniformly by the monitoring room. Each on-site operator can view the monitoring data of their respective sites, but does not have control permissions.

⑩ The data monitoring cycle can be adjusted, and the data is automatically stored, analyzed, and arranged according to the monitoring cycle to ensure data traceability.

(1) System Introduction

The configuration of this system is divided into the following parts.

UIOTC Cloud Platform Data Viewing and User Interface, as shown in Figure 3.2.5.

Figure 3.2.5 UIOTC Cloud Platform Display Page

① MD9000 Wireless Gateway (4G/LoRaWAN/CAT1/NBIoT): Sends data from multiple on-site collection modules, MD9103E, to the cloud server via 4G/LoRaWAN/CAT1/NBIoT.

② Wireless Acquisition Module MD9103E Function: The MD9103E module serves as the acquisition center to upload the data from digital temperature sensors on the measuring cable and two isolated switch inputs (dry contacts) (which can connect sensors such as smoke alarms, groundwater, water level alarms, anti-theft, and intrusion detection, etc.) to the MD9000 wireless gateway. It supports long-term battery power supply (nominal 3 years or more).

MD98XXE Customized Sensor Spacing High-strength Waterproof and Anti-interference Type Temperature Measurement Cable. Measurement accuracy is better than ±0.2 ℃ (actual test data is better than ±0.1 ℃), resolution is up to 13 bits, i.e., 0.03125 ℃.

Note: This plan is compatible with the MD9102E wireless data acquisition terminal for various environmental data and standard signals, making it easy to add standard analog quantities such as humidity probes, gas sensors, and current monitoring.

(2) System Structure

The system adopts a fully digital network structure, which improves the anti-interference ability of the entire system. Each MD9000 wireless gateway can support thousands of 9103E. Each MD9103E module can support 20 digital temperature sensors. The data communication of the system uses CRC8 error correction to ensure the reliable operation of the system in harsh environments.

System Topology Diagram as shown in Figure 3.2.6.

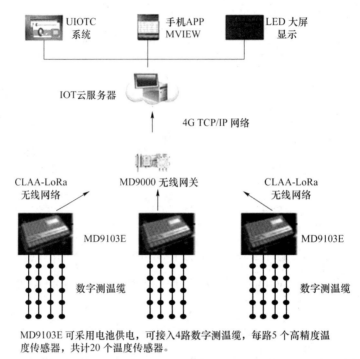

Figure 3.2.6　Topological Structure Diagram of Hydration Heat Monitoring System

The system hardware configuration list is shown in Figure 3.2.7.

The IoT wireless high-precision concrete hydration heat cloud monitoring platform solution is an advanced and practical solution that can provide engineers with real-time and accurate temperature monitoring data and analysis reports, ensuring the safety and stability of engineering projects.

3. Implementation Plan

The basic process of implementation includes:

① Determining monitoring objectives: Clearly define the purpose of monitoring, such as monitoring the heat of hydration in large-volume concrete to provide a basis for construction.

② Determining the monitoring plan: Based on specific circumstances, select appropriate monitoring methods and techniques, such as using temperature sensors for single-point or continuous monitoring, and determine the frequency and duration of monitoring.

CLAA-LoRa混凝土水化热监测系统配置清单					
设备名称	型号规格	单位	数量	备注
环境监测云平台系统	UIOTC	套	1	Wed版监控软件,实现功能:环境数据(温湿度等)、网页视频、电子地图仪表显示、曲线显示、手控设备、自控设备、门禁管理、视频管理、实景组态、2D扩散呈现功能、UPS监测、精密空调监测等
CLAA-LoRa网关	MD9000	台	1	IWG:无线物联网网关(Iot Wireless Gateway0),实现LoRa物理层功能,支持16个信道并发,470-510 M宽范围跳频抗干扰,接入各类LoRa应用节点,实现链路安全、压缩等功能,支持3G/4G及有线Backhaul链路,支持蓝牙就近无线配置管理,支持IWG间高精度时钟同步,为实现低成本/低功耗LoRa定位打下基础,支持市电、太阳能供电,抱杆、挂墙等多种部署模式,根据应用场景不同支持IP65或IP68防护等级两种型号,−40到70℃的宽温,抗恶劣环境室外型工业级设备
无线采集模块	MD9103E	台	若干	MD9103E模块作为采集中心将测量线缆上的数字化传感器的数据及两路隔离型开关量(干接点)输入(可接烟感、地面积水、沟内水位报警、井盖开盖监测等传感器)信号。可电池供电
订制高强度数字测温缆	MD98XXE	根	若干	每个MD9103E模块可接4路数字测温缆,最多20个数字温度传感器

Figure 3.2.7 Configuration List of Hydration Heat Monitoring System

③ Preparing monitoring equipment: Prepare the corresponding monitoring equipment according to the selected monitoring plan, such as temperature sensors, data loggers, and analysis software.

④ Establishing a monitoring system: Install and debug the prepared equipment according to the requirements of the plan, and establish a stable monitoring system.

⑤ Monitoring: During the designated monitoring period, real-time monitoring of the temperature of the concrete body is conducted, and data are recorded.

⑥ Data Processing and Analysis: The collected data are processed and analyzed, including temperature change trends, hydration heat peaks, etc., in order to understand the hydration heat situation of the concrete body.

⑦ Evaluation and Feedback: Based on the monitoring results, the hydration heat situation of the concrete body is evaluated, and corresponding suggestions and measures are proposed, and feedback is given to the construction party.

It should be noted that when implementing the concrete body hydration heat monitoring system, the accuracy and stability of the monitoring equipment should be ensured, and relevant safety regulations and operating procedures should be followed to ensure construction safety.

4. Scheme Verification

The main steps of scheme verification include:

① Determining the verification objectives: Clearly define the objectives of the verification, such as verifying the accuracy and stability of the monitoring system.

② Preparing verification data: Prepare hydration heat data of the concrete body for verification, which can be actual data obtained through other methods or simulated data.

③ Implementing the monitoring system: Establish and implement the monitoring system according to the implementation plan, and conduct real-time monitoring.

④ Collecting monitoring data: Collect temperature data of the concrete body within the designated monitoring time and record it.

⑤ Data processing and analysis: Process and analyze the collected data, compare it with the verification data, and evaluate the accuracy and stability of the monitoring system.

⑥ Evaluation and Feedback: Based on the comparison results, evaluate the monitoring system. If there are significant errors or insufficient stability, adjustments and improvements to the system are necessary.

Here, we mainly experimentally verify the configuration of the gateway, terminal, cloud platform, and client configuration. There are many products available on the market regarding the LoRaWAN system. In this solution, we have chosen the gateway model RHF2S024. The cloud platform uses the open-source LoRaWAN NS server, the application platform uses Micronode Technology's IoT cloud, the collection module uses MD9103 and MD8877, and the configurable terminal module selects the S78S LoRa module from Acsip Technology.

Experiment 1 Network Server Configuration

Understand the role of NS in the LoRaWAN system and master the configuration process of the gateway and terminal on the NS.

(1) Experimental Principle

The NS used in this experiment, LoRaWAN-server, is an open-source LoRaWAN server that integrates NS and AS functions to facilitate the connection between devices and application developers.

In this experiment, it is necessary to configure the Gateway, Network, Groups,

Handlers, Profiles, Commissioned, and Connectors on the NS server to ensure that the gateway and terminal devices can access the network and transmit data.

(2) Experiment Steps

① Log in to the NS with the username, admin and password: admin. The main interface is shown as Figure 3.2.8.

Figure 3.2.8　NS Main Page

② To configure the gateway, click Infrastructure → Gateways → Create, as shown in Figure 3.2.9.

Figure 3.2.9　NS Gateway Configuration Page

Note：The MAC address should be filled in with the actual MAC address of the gateway being used.

③ To configure the network，click Infrastructure → Networks → Create，as shown in Figure 3.2.10～Figure 3.2.12.

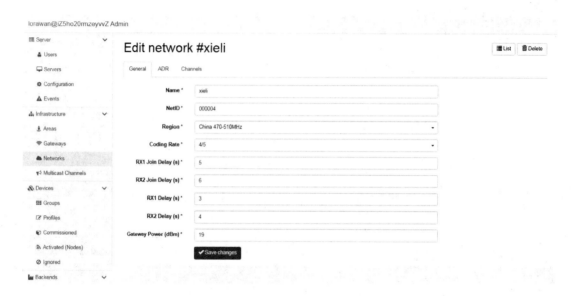

Figure 3.2.10　NS Network General Configuration Page

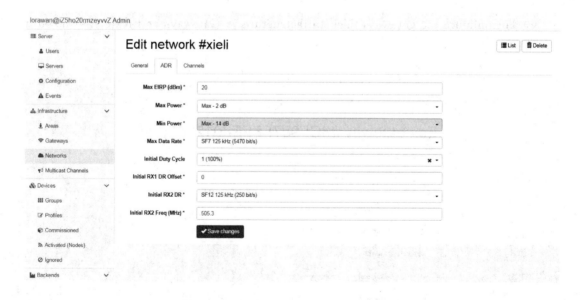

Figure 3.2.11　NS Network ADR Configuration Page

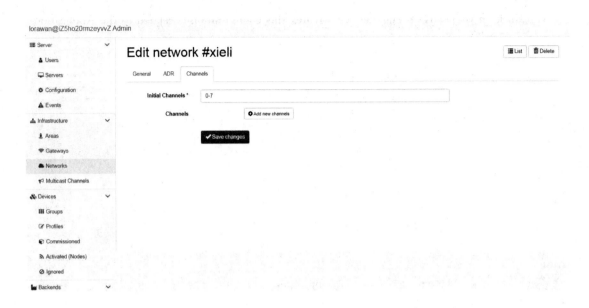

Figure 3.2.12　NS Netowrk Channels Configuration Page

④ To configure equipment information, you need to configure Groups, Handlers (shared by a group; students do not need to configure), Profiles, and Commissioned information, as shown in Figure 3.2.13~Figure 3.2.17.

a. Click on Devices → Groups → Create.

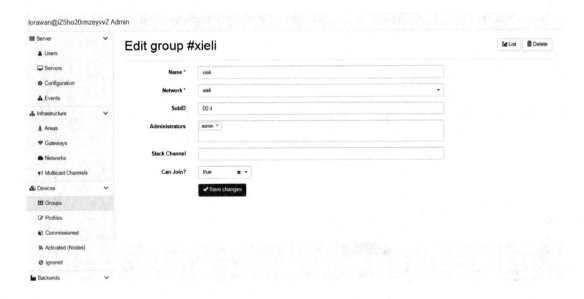

Figure 3.2.13　NS Groups Configuration Page

b. Click on Backends → Handlers → Create, this configuration is related to the Application.

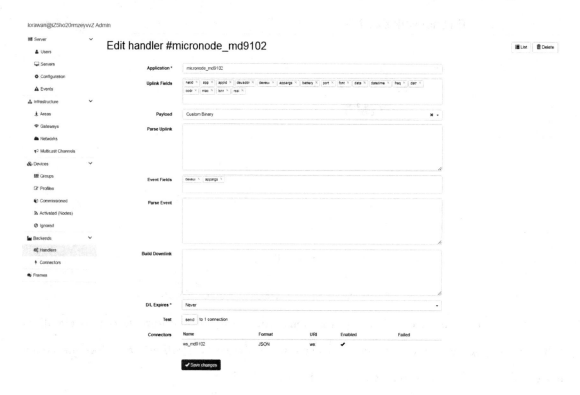

Figure 3.2.14　NS Handler Configuration Page

c. Click on Devices → Profiles → Create, select the already configured Groups and Application.

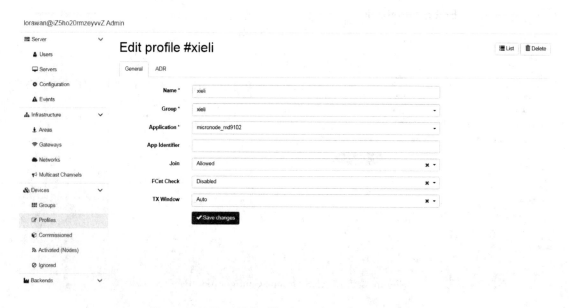

Figure 3.2.15　NS Profile General Configuration Page

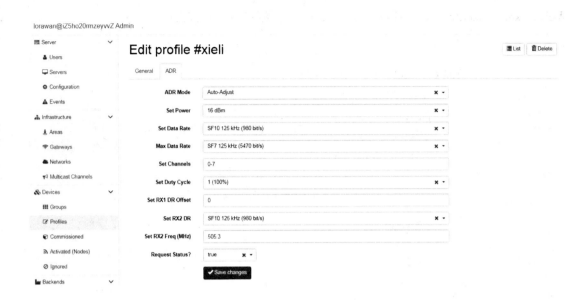

Figure 3.2.16   NS Profile ADR Configuration Page

d. Click on Devices → Commissioned → Create, fill in the actual device's DevEUI, AppEUI, and AppKey, otherwise, the device will not be able to connect to the network properly.

Figure 3.2.17   NS Device Configuration Page

⑤ To configure the connector (multiple devices sharing a group), click on Backends → Connectors → Create, the Connector is used to communicate with the application server, and

WebSocket is used here, as shown in Figure 3.2.18~Figure 3.2.19.

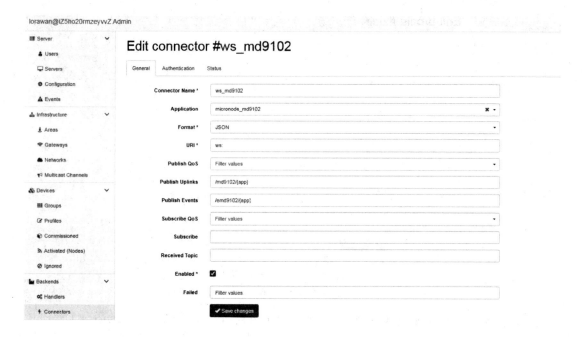

Figure 3.2.18　NS Connector General Configuration Page

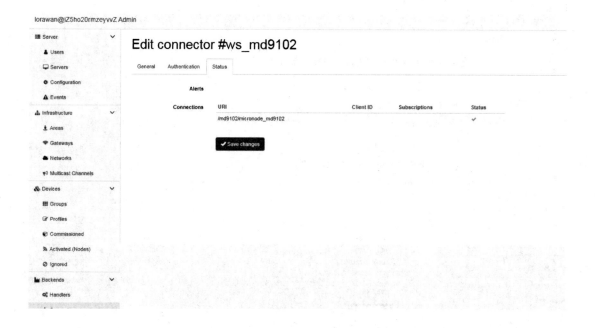

Figure 3.2.19　NS Connector Status View Page

⑥ You can view the configured gateway and device status on the main interface.

⑦ Click on Frames to view the data stream uploaded and downloaded by each device, as shown in Figure 3.2.20.

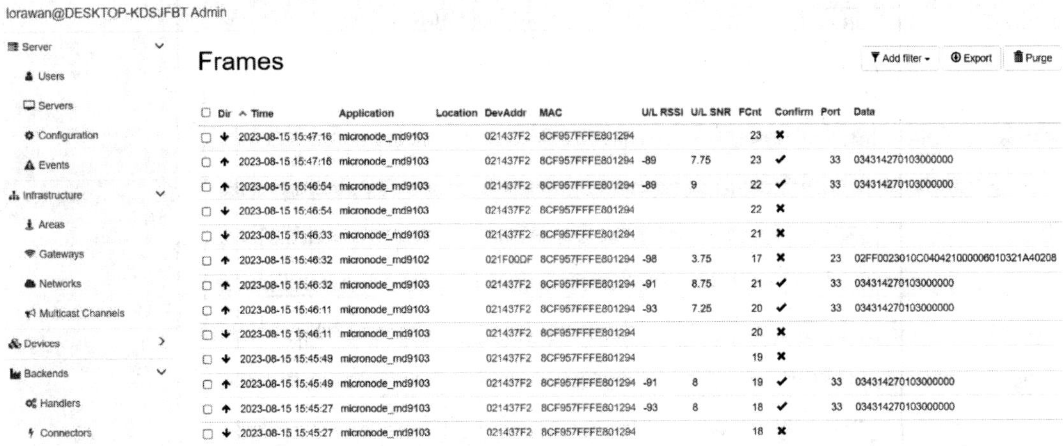

Figure 3.2.20　NS Stream View Page

Experiment 2　Multipoint Temperature Acquisition Module MD9103

Understand the working principle of the MD9103 and master the usage of the MD9103 and MD8877.

(1) Experimental Principle

MD9103 is a LoRa multipoint digital temperature sensor acquisition module that can be connected to MD8877 digital temperature sensor. It has 4 sensor input channels and 2 optically isolated switch input channels. It is powered by batteries (an external power supply mode is optional). The MD9103 starts data acquisition at regular intervals and uploads the temperature measurement points and switch data from the field to the LoRa network, allowing for the construction of various sizes of on-site monitoring application systems.

The appearance of MD9103 is shown in the Figure 3.2.21.

Explanation:

BT1—18650 battery compartment, two batteries can be installed simultaneously and used in parallel. Note: Pay attention to the polarity marking when installing the batteries!

CN1—Optional external 5 V DC power input, cannot be used simultaneously with battery mode. 1—+5 V, 2—GND.

JB —Spare external battery interface (Note: Only use when there is no battery in the BT1 battery compartment!)

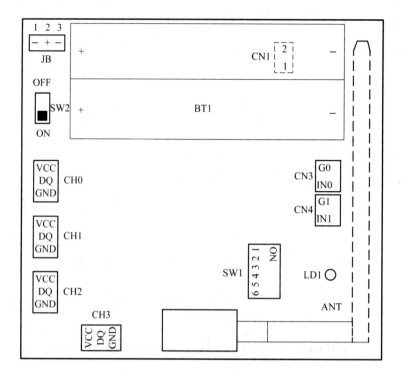

Figure 3.2.21 MD9103 Appearance Diagram

Battery voltage requirement: 3.6~4.2 V.

SW2—Battery power switch, ON—On, OFF—Off. Only used in battery mode.

CH0-CH3—External channels for digital temperature sensors.

CN3, CN4—Switch input channels.

SW1—Module operating mode setting.

LD1—LED indicator light.

ANT—LoRa antenna.

After the module is powered on and obtains the correct sensor parameters (readstorage/automatic search), it will start the CLAA-LoRa network joining process. During this stage, the MD9103 module will attempt to join the CLAA network, which takes about 10~90 s. At this time, the LED light LD1 will flash rapidly, and once LD1 stops flashing, it means that the module has successfully joined the network. After that, the module will enter the working process.

(2) Experiment Steps

① Correctly connect the antenna and power supply of the MD9103 to ensure that the internal dip switch is set correctly.

② Connect MD8877 to the external sensor channel of the MD9103.

③ Perform steps ③~⑤ in Experiment 2 to configure the MD9103 on the NS server.

④ Power on the MD9103, observe the change of indicator light LD1, and check the terminal status and uploaded data on the NS server.

Experiment 3　Configuration and Use of UIOTC (Your IOT Cloud)

Understand the role of UIOTC in the LoRaWAN system, master the configuration process of the terminal on the platform, and learn how to use the platform.

1) Experimental Principle

The UIOTC used in this experiment is an environmental monitoring cloud platform that can meet the cloud-based environmental monitoring needs of a single project, as well as serve as a general monitoring cloud platform to meet the real-time environmental monitoring needs of multiple users. With the support of data acquisition devices, the system can centrally implement cross-regional and cross-gateway environmental monitoring. It can list and view real-time and historical data and has practical functions such as real-time alarm, chart display, and real scene configuration. It can also be combined with the mobile application APP to achieve comprehensive industrial environmental monitoring.

For this experiment, it is necessary to configure the relevant information of each group device on the UIOTC to ensure the display and alarm of uploaded data.

2) Experiment Steps

(1) Log in to the UIOTC. The main page is shown in Figure 3.2.22.

Figure 3.2.22　UIOTC Login Page

(2) After logging in, you can see the user's basic information. If you need to modify permissions, click "Modify", as shown in Figure 3.2.23~Figure 3.2.24.

Figure 3.2.23   UIOTC User Management Page 1

If you need to add a new user, click "Create User".

Figure 3.2.24   UIOTC User Management Page 2

(3) Perform configuration management and operate project management, device group management, and device management separately, as shown in Figure 3.2.25~Figure 3.2.33.

① Project management: Click "Configuration Management" → "Project Management" to modify project-related information. (The project used this time has been created by the administrator, and students can view the information, but it is not recommended to make modifications.)

Figure 3.2.25   UIOTC Project Management Configuration Page

② Device group management: Click "Configuration Management"→ "Device Group Management". On this page, you can create new groups or modify existing group information.

Figure 3.2.26　UIOTC Group Management Configuration Page 1

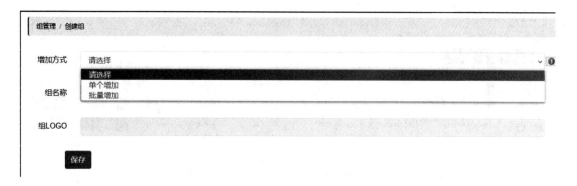

Figure 3.2.27　UIOTC Group Management Configuration Page 2

③ Device Management. On this page, you can add a new device or add, modify, and delete sub-sensors for an existing device.

a. Click on Add Device.

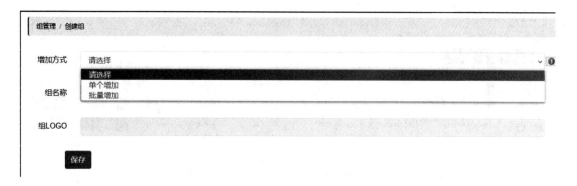

Figure 3.2.28　UIOTC Device Management Configuration Page 1

b. Fill in the device information. The device name must be unique, and the device address should be filled in with the DEVEUI configured on NS.

Figure 3.2.29　UIOTC Device Management Configuration Page 2

c. To add a sub-sensor, click on Sensor Configuration.

Figure 3.2.30　UIOTC Device Management Configuration Page 3

d. Taking the addition of 8877 under MD9103 as an example, fill in the device number with the sensor address connected to MD9103.

Figure 3.2.31　UIOTC Device Management Configuration Page 4

e. Modifications and deletions can also be made to existing sub-sensors.

④ For data queries, you can view real-time data, real-time curves, historical data, alarm information, and realistic display. You can also export the queried data results, as shown in Figure 3.2.34 and Figure 3.2.35.

Figure 3.2.32 UIOTC Device Management Configuration Page 5

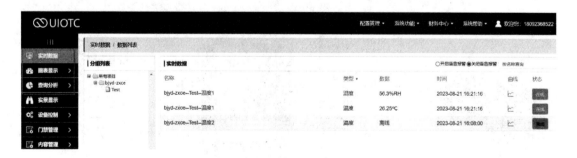

Figure 3.2.33 UIOTC Device Management Configuration Page 6

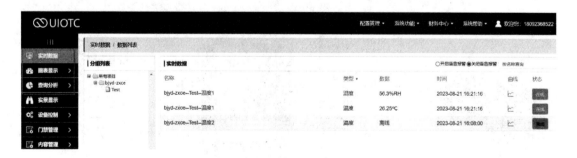

Figure 3.2.34 UIOTC Data View Page

Figure 3.2.35　UIOTC Realistic Display Page

Experiment 4　Configure S78S Module to Access Cloud Platform

Understand the role of terminal modules in the LoRaWAN system, and master the configuration methods and key commands of the modules.

(1) Experimental Principle

The S78S module integrates ATmega328P MCU and LoRa module, provides GPIO/I2C/ADC/UART interfaces, can connect multiple sensors and power modules externally, and supports three types of instructions: SIP command instruction set, MAC command instruction set, and RF command instruction set. The appearance and connection diagram of the module are shown in Figure 3.2.36.

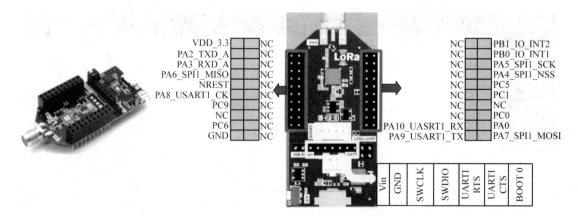

Figure 3.2.36　Appearance and Pin Diagram of S78S Module

By setting the channel center frequency, DEVEUI, APPEUI, and APPKEY of the module to the gateway through the AT commands, the module can access network and upload data.

(2) Experiment Steps

① Connect the S78S module to the PC through a USB port, as shown in Figure 3.2.37.

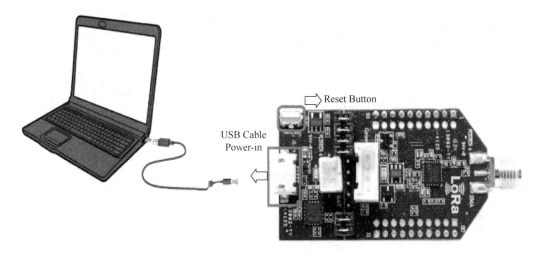

Figure 3.2.37　S78S Module Connected to PC

② Open software "Termite" and set the transmission baud rate to 115 200, as shown in Figure 3.2.38.

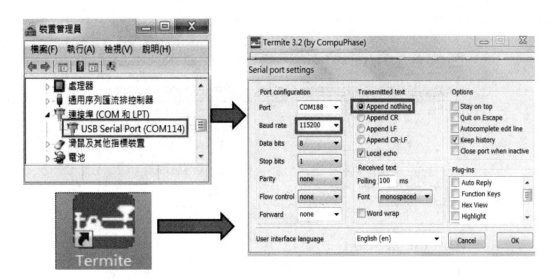

Figure 3.2.38　Set Transmission Parameters in "Termite"

③ Initialize the module by entering the command "sipfactory_reset" to set wireless transmission-related parameters to default values. Enter the command "Mac save" to save the configuration parameters to the cache. Enter the command "sip reset" to restart the module firmware. After each parameter modification is completed, a firmware restart is required, as shown in Figure 3.2.39.

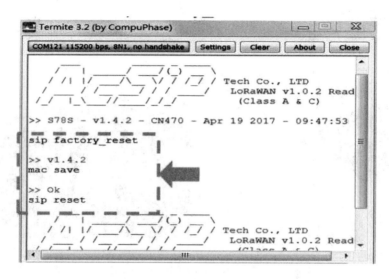

Figure 3.2.39 Restart the Firmware of the S78S Module

④ Enter the commands "sipget_ver" and "sip get_hw_model" to obtain the current firmware version and name, as shown in Figure 3.2.40.

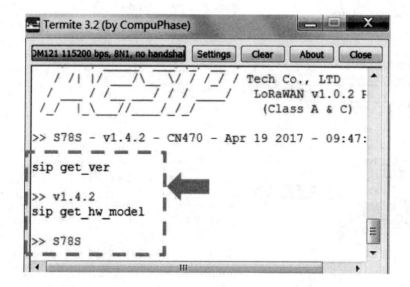

Figure 3.2.40 Obtain Module Firmware Version and Name

⑤ Enter the command "macset_ch-freq<ChannelId><Frequency>" to set the center frequency points of each channel, as shown in Figure 3.2.41.

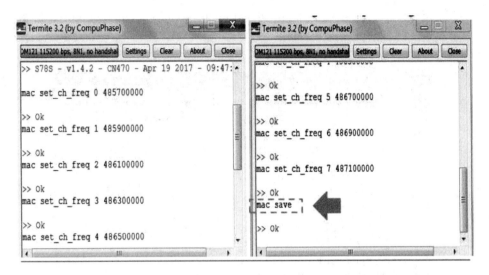

Figure 3.2.41　Set Center Frequency Points for Each Channel

⑥ Enter the command "macget_ch_para<ChannelId>" to obtain the uplink frequency point, minimum data rate, maximum data rate, frequency band number, and downlink frequency point for each channel, as shown in Figure 3.2.42.

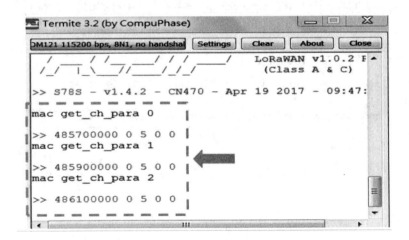

Figure 3.2.42　Obtain Channel Information

⑦ The module is connected to the network through the ABP method. First, confirm that the relevant parameters for module connection are the default values, and then set the parameters to the planned values. Confirm the default parameters by entering commands such as "mac_get_devaddr", "mac_newkskey", and "mac_get-appskey"; Enter the commands "mac set_devaddr<DevAddr>" "mac set_cwkskey<NwkSKey>" "mac set_appskey<AppSKey>" to set the planned parameters, as shown in Figures 3.2.43 and Figure 3.2.44.

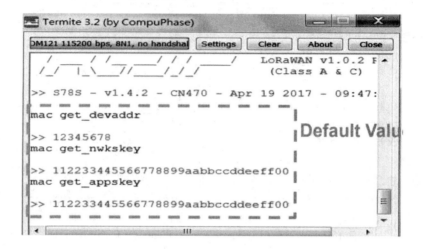

Figure 3.2.43　Obtain Network Access Parameters

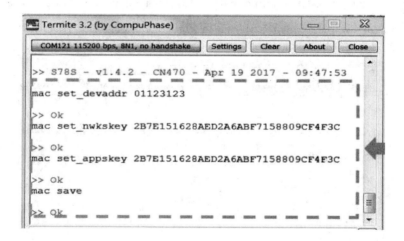

Figure 3.2.44　Set Network Access Parameters

⑧ Enter the command "Mac join<Mode>" to set the module's network access method, taking ABP network access as an example, as shown in Figure 3.2.45.

⑨ If the module is connected to the network through OTAA, simply enter the commands "mac set_deveui<DevEUI>" "mac set_appeui< AppEUI >" "mac set_appkey <AppKey>" to set the parameters. Enter the command "mac join otaa" to join the network, as shown in Figure 3.2.46. These three parameters need to be consistent with the ones filled in on NS, as shown in Figure 3.2.47.

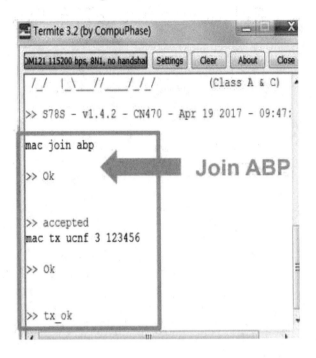

Figure 3.2.45 Set Network Access Method

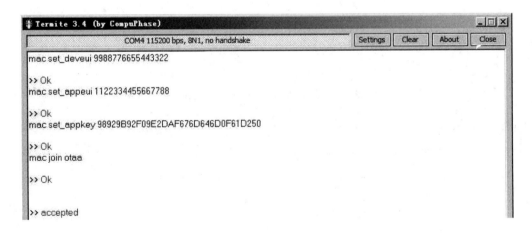

Figure 3.2.46 Join the Network Through OTAA

⑩ After the module is successfully connected to the network, the corresponding network access information and the data uploaded by the module can be seen on NS, as shown in Figure 3.2.48 and Figure 3.2.49.

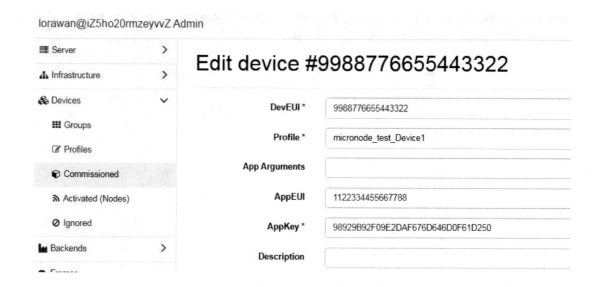

Figure 3.2.47　Three Parameters on NS

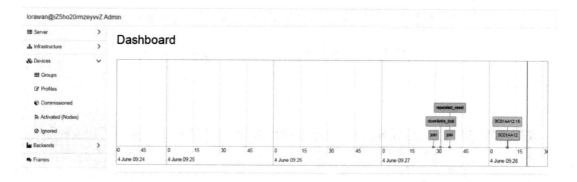

Figure 3.2.48　Network Access Information on NS Main Page

Figure 3.2.49　Module Upload Data on NS Stream View Page

5. Question and Thinking

Question 1: Please list the various types of data that need to be monitored by the concrete hydration heat monitoring system and explain the necessity of monitoring each type of data.

Question 2：How to confirm that the module S78S network access parameter settings are consistent with the gateway?

Question 3：If S78S is connected to the network through OTAA，what parameters need to be set?

3.3　乡村振兴

3.3.1　智慧仓储

智慧仓储是当今仓储行业变革的重要方向，主要是为解决传统仓储行业人力成本高，仓储效率低，仓储管理复杂等问题，而结合现代科技设计的一套仓库管理系统。

所谓智慧仓储是指在仓储管理业务流程再造基础上，利用 RFID（射频识别）、网络通信、信息系统应用等信息化技术及先进的管理方法，实现入库、出库、盘库、移库管理的信息自动抓取、自动识别、自动预警及智能管理功能，以降低仓储成本，提高仓储效率，提升仓储智慧管理能力。

在全球经济与科学技术的双重推动下，我国的仓储管理已进入高速发展期，物流仓储也由初期简单的自动化阶段开始向数字化智慧仓储跨越。

智慧仓储功能包括以下 4 个方面。

① 仓储信息自动抓取。仓储信息自动抓取功能是指对贴有电子标签的货物、库位、库架信息自动抓取，包括货物属性、库位及库架分类等，无须通过人工一一辨认。

② 仓储信息自动识别。仓储信息自动识别功能是通过与后台服务器连接，在自动抓取信息基础上完成的，从而实现信息自动识别，快速验证出入库货物信息、库内货物正确堆放信息等功能。

③ 仓储信息自动预警。仓储信息自动预警功能是通过信息系统程序设定，对潜在的问题货物进行预警，以便于提前采取措施应对。

④ 仓储信息智能管理。仓储信息智能管理功能可以自动生成各类单据，是为供应链决策提供实时信息的功能模块。

智能仓储的工作单元包括软件单元、硬件单元、网络单元、管理单元四大部分。

① 软件单元。智慧仓储的软件单元为智慧仓储管理信息系统，主要包括基本信息管理模

块、货物出入库管理模块、货物盘库管理模块、标签、阅读器管理模块、货物预警模块与智慧仓储管理模块七大模块。

② 硬件单元。硬件单元包括 RFID 电子标签、读写器、阅读器、RFID 电子标签打印机、服务器、终端、仓库基础设施等。

③ 网络单元。网络单元由计算机有线网络及无线网络组成,其中无线网络主要指 Wi-Fi 及 GPRS(通用分组无线服务)两类。

④ 管理单元。管理单元是指一套基于智能仓储的管理业务流程与规范,主要包括出入库、盘库、移库作业流程及相应的规范要求。

1. 场景分析

进入 21 世纪以后,企业经营的内外环境的前景都存在一系列不确定因素,全球化带来的竞争国际化导致企业的压力进一步加重。烟草行业面临着同样的境遇,长期滞后的薄弱的管理基础已难以支撑企业继续快速发展。对于烟草行业来说,推行精细化管理是不二之选。

在烟草整个生产周期中有非常多的环节:烟叶种植、打叶复烤、烟草醇化、制丝以及卷包生产等。一般在烟叶种植前一年,农户就与当地烟草公司进行合同签订,确定烟叶品种、收购量。在烟叶成熟后,农户将其送往复烤厂。在复烤厂,烟叶按照批次进行麻包堆垛,再由专业质检人员将烟叶进行鉴定分级。挑选完的烟叶开始按照一定工序上不同的生产线进行复烤操作,等待所有工序生产下线,这时就得到了真正意义上的第一个批次——复烤批。当复烤批产生之后,也就形成了卷烟厂中常见的原烟。复烤出来的原烟是无法直接用于卷烟生产的,需要在醇化完成后再运到各地卷烟厂进行卷烟的制作(醇化过程一般持续 3 年)。所谓醇化就是将刚复烤完成的原烟运输到一个个醇化仓库,这些仓库保持在一定的温湿度,从而将烟叶中一些有害成分进行发酵并挥发。最后送到卷烟厂生产销售。

本方案主要是针对烟草醇化阶段,影响卷烟品质的好坏的一个重要因素是烟叶自然发酵的好坏。在烟叶自然发酵的仓储过程中,温度和湿度是事关发酵效果的两个重要指标。在温度方面,过高的温度会引发烟叶自燃、碳化变质,过低的温度又会妨碍烟叶的发酵呼吸。在湿度方面,太高的湿度是引起霉变的重要原因,太低的湿度又将失去自然发酵的意义。据统计经验表明:在温度范围为 20~25 ℃,湿度范围为 65%RH~75%RH 时,仓储发酵的效果最好,能获得最佳的卷烟品质。

在烟草醇化阶段,传统烟草企业对仓库温度、湿度采用的监控手段还是通过壁挂式温湿度表,也就是说,操作人员主要依靠对壁挂式温湿度表的读数及个人经验来开闭仓库门窗,从而对仓库温湿度进行调节。在这个过程中,读数往往误差大,不能准确反映烟叶垛内部的温度、湿度,因此易导致烟草品质差,不理想。

针对以上问题,需要对烟草醇化库设计一套智能的监控系统,监控系统需要实现以下功能。

① 异常报警系统:系统出现异常后能发出多媒体语音报警,同时软件界面弹出报警提示,并以电话语音或手机短信方式发出告警;支持手机和电脑通过 Web 方式进行远程浏览、管理。

② 智能采集模块:采集传感器信号,监测烟叶垛内部温度数据、仓内温湿度数据、仓间磷化氢等数据;输出控制信号,控制智能门窗、智能通风、空调、除湿机、加湿器等设备。

③ 视频监控系统:通过接入的视频监控设备,对仓内、仓库周围状况及人员进出实时监控,一旦有异常事件发生,监控系统自动弹出现场图像画面,即时录像并进行报警提示和处理。

④ 安全防护系统:门禁系统实现库房大门的开关控制和进出人员记录等;电子围栏系统实现仓库周围的非法侵入报警;水浸传感器实现地面积水报警;消防辅助系统,即在重点区域增加烟雾探测器,缩短消防响应时间。

2. 项目方案设计

在这个前提下,北京微点科技凭借因 10 多年服务粮食行业而沉淀得来的丰富的烟草醇化库数字化产品设计和生产经验,综合了先进的仓储管理技术、电子技术、计算机网络技术、通信技术及防腐防雷技术,为新一代烟草醇化库量身定做了一套智能监控系统。

由于新一代烟草醇化库涉及的系统很庞大,我们以其中的智能通风系统为例,来介绍北京微点科技对该子系统的设计。

该子系统由智能控制主机、电动/手动一体式执行机构等主要部分构成,可支持在恶劣气候下,进行应急短信远程操作或手机客户端 APP 应急调节,实现了仓库在无人值守的状态下的智能控制。

烟草醇化智能监控系统结构如图 3.3.1 所示。

智能通风系统能与现有的烟叶醇化测控系统连接,并利用计算机测控系统对烟叶垛温度、仓内温湿度和气象条件进行实时数据监测。依据《储粮机械通风技术规程》、储粮具体情况、地理环境条件、仓库特点等,建立专家模型数据库,进行综合、快速地智能分析,准确判断允许通风的各项条件,捕捉最佳时机以自动控制轴流风机、离心风机等通风设备、自动通风口、自动通风窗的开闭等。根据通风目的(排积热、降温等),进行排积热通风、降温通风等。

烟草醇化智能通风系统结构如图 3.3.2 所示。

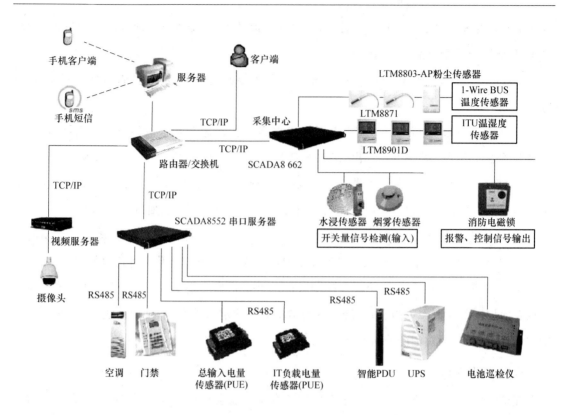

图 3.3.1 烟草醇化智能监控系统结构

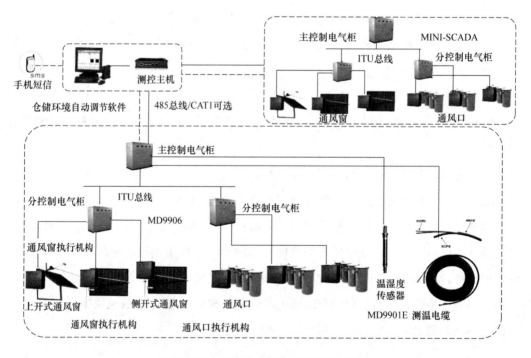

图 3.3.2 烟草醇化智能通风系统结构

在该智能通风系统中,测控主机选用高性能、无风扇、宽温、嵌入式工控机,其适合长时间工作,内置CAT1模块,可支持4G/5G数据通信或短信收发;内置仓储环境自动调节软件能对烟叶垛内部温度、仓内温湿度和仓间磷化氢浓度进行实时数据监测,根据储粮具体情况、地理环境条件、仓库特点等,建立数学模型,自动控制通风设备及设施的开关或启停,从而实现最佳通风效果。

① 主控制电气柜。MINI-SCADA专为智能通风系统设计,可采集烟草醇化库所在区域的仓内、仓外温湿度数据,控制多个通风窗执行机构和通风口执行机构,同时可读取通风窗关闭状态。硬件配备看门狗,可在无人值守场合下长时间无故障运行。

MINI-SCADA和烟叶醇化数字温度传感器均采用自有知识产权的ITU总线技术,既可以就近安装,方便布线,也兼顾了成本系统可靠性和便捷性。

ITU总线产品专为产品客户设计。ITU总线是传输距离小于300 m的设备总线(Micro LAN),是方便现场分散布线的一条可靠、经济的总线。ITU总线几乎是最方便的产品扩展单元,它带来了前所未有的产品设计灵活性,同时又兼顾了现场的安全性与成本的经济性。

自动通风窗选用烟草醇化库专用密封保温窗,是根据智能通风系统要求开发的专用产品。选用优质型钢制作窗框和骨架,面板选用彩板或镀锌板,内腔填充聚乙烯泡沫板。窗的开关是通过通风控制器或操控手柄,带动连杆机构完成,弹簧销结构简单,锁紧可靠。内置模块采用计算机单片及智能控制软件直接接收计算机或工业仪表输出标准控制信号,可实现电动阀门开度的智能控制和精确定位。

自动通风口选用烟草醇化库专用通风口设计和防水式设计,是根据智能通风系统要求开发的专用产品。内置模块采用计算机单片及智能控制软件直接接收计算机或工业仪表输出标准控制信号,可实现电动阀门开度的智能控制和精确定位。

② 分控制电气柜。MD9906专为智能通风系统设计,可无线接收上位机发出的控制信号,控制多个通风窗,从而启动或停止通风窗。硬件配备看门狗,可作为无人值守现场下长时间无故障运行的坚强保障。支持自有知识产权的ITU总线,一条ITU总线上可传输多个MD9906或其他ITU设备信号。

③ 温湿度传感器。MD9901E专为智能通风系统设计,适合测量仓内、仓外温湿度数据。数字式温湿度一体化探头,烟草醇化库专用烧结铜网防护设计,支持自有知识产权的ITU总线设计,一条ITU总线上可传输多个MD9901E或其他ITU设备信号,通过主控制电气柜将数据上传给测控主机。采用POE方式供电,有测控主机为其远端供电,无须现场供电。

智能通风系统软件提供了丰富实用的功能,不仅能够给用户提供一个远程控制通风设备的操作平台,还能够进行自动把握通风过程中的烟叶状态,根据用户预设好的通风条件,自动逻辑判断是否执行和终止通风。智能通风系统提供了降温通风、降水通风、调质通风和排热换气通风四种通风模式,根据用户的通风目的自行执行通风烟叶醇化的逻辑判断,自动捕捉最佳的通风时机,避免低效通风、无效通风和有害通风现象的发生,提高通风效率。该系统可成功解决常规机械通风中通风条件难于判断、控制等问题,实现通风智能化、自动化,以及烟叶醇化的智能化分析判断和预警预报,确保仓储管理和技术人员准确了解烟叶醇化变化和及时发现隐患。

在智能通风软件系统,配置包含对降温通风、降水通风、调质通风、排热换气通风四种通风模式的参数设置界面,设定完成之后进行保存即可。参数设置界面如图3.3.3～图3.3.5所示。

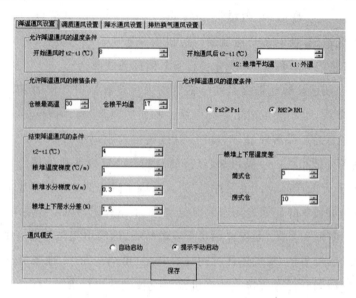

图3.3.3 烟草醇化智能通风系统参数设置界面图1

图3.3.4 烟草醇化智能通风系统参数设置界面图2

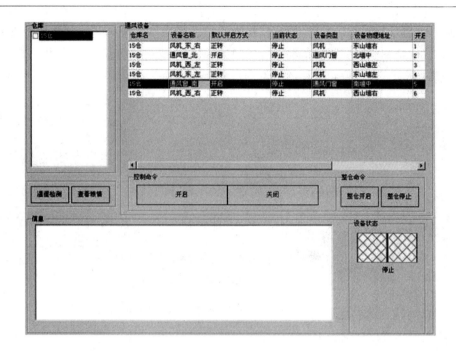

图 3.3.5　烟草醇化智能通风系统参数设置界面图 3

在工作模式配置界面，用户可设定通风类型和烟叶醇化检测的间隔时间，在通风模式中还可以选择全天候、白天或夜晚通风。

该系统具有人性化的远程操作面板，可以根据用户在当前仓库列表中选择的仓库，执行整仓命令，省时省力。

汇总各仓的风网设置信息和分仓设备信息，以及本次通风前后的烟叶醇化变化信息，可以通过选择仓和通风日期进行查看和打印操作。如图 3.3.6 所示。

储粮机械通风工作记录卡

粮库名称:	15仓	仓号:	1	通风次数:	
仓型:	房式仓	尺寸:			
粮食种类:		数量:	0		
通风目的:	降温通风	粮堆高度:			
风机类型及型号:	风机_轴流风机	通风方式(吸/压):			
风机总阻力:	0.2	气流方向(上/下行):	向下		
总风量:	0	单位通风量:			
风机总功率 :	0				
风网布置	风网类型	地槽通风	通风开始时间:	2007-7-8 9:47:09	
	空气通径比	0.3	通风结束时间	2007-7-8 9:49:41	
	孔板开孔率:	0.15	通风累计时间	0天0小时2 分钟	
大气温度:		大气湿度:			
通风时粮食参数		最高温	最低温	平均温	
粮仓温度	通风前	0	0	0	
	通风后	0	0	0	
粮食水分	通风前	0	通风后	0	
通风效果:	无效通风				
操作员签章:		负责人签章:			
备注					

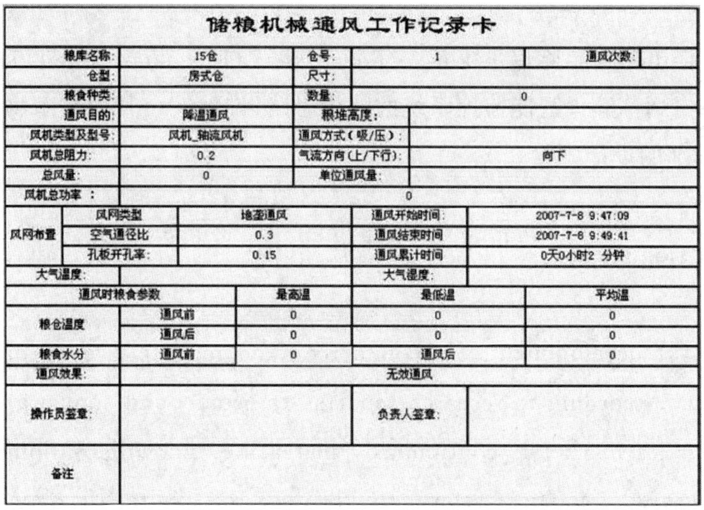

图 3.3.6　烟草醇化智能通风系统工作记录卡

此外，该系统还可以给出科学分析，选择醇化仓库和通风时间以及烟叶醇化参数类型，查看在通风过程中烟叶温度变化的情况，并将这些数据转换成柱状图或曲线图，从而将烟叶醇化的变化情况以更加直观的形式呈现给用户，如图3.3.7所示。

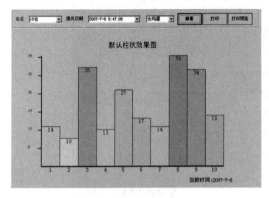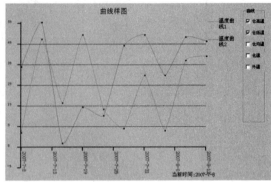

图 3.3.7 烟草醇化智能通风系统数据查看页面

该监控系统具有以下优势：

① 数据化：烟叶醇化过程温湿度、磷化氢浓度实时监测。

② 立体化：三维分解、立体监测烟叶垛各物理指标。

③ 自动化：电脑自动采集和记录，并根据所得数据进行智能判断是否开启门窗进行通风，或控制设备调节温湿度。

④ 标准化：建立一套企业内部烟叶醇化管理的环境控制规范，通过长期的、密集的三维数据不断优化烟叶醇化技术水平，并通过自动化体系快速应用到企业各仓库，提高整体醇化水平。

3. 问题与思考

问题1：本节提出的智能监控系统是否有系统缺陷？如有，请提出完善方案。

问题2：请根据烟草醇化智能监控系统的分析思路，设计一个其他类型物品的仓储智能监控系统，并给出详细的设计方案。

3.3.2 Digital Logistics

With the rapid development of e-commerce, the market size of digital logistics is expanding rapidly. According to research reports, the compound annual growth rate of the global digital logistics market is continuously increasing and will continue to grow in the coming years. Digital logistics refers to the use of digital technology to digitize, intelligentize, and networking various aspects of the traditional logistics industry, thereby improving logistics efficiency, reducing costs, enhancing service quality, and optimizing supply chain management. Digital logistics is characterized by full digitization, full visualization, and full intelligence, and it is an inevitable trend for the logistics industry to transform into digital and intelligent.

The core technologies of digital logistics include the Internet of Things, big data, cloud computing, artificial intelligence, etc. The continuous innovation and development of these technologies provide strong support for digital logistics. For example, IoT technology can track the location and status of goods in real-time, while big data and cloud computing can optimize logistics routes and improve delivery efficiency. Intelligent logistics systems can automatically complete tasks such as sorting, packaging, and delivery of goods through artificial intelligence technology, and automated equipment can reduce manual labor and improve work efficiency.

Traditional logistics companies are accelerating their digital transformation, while emerging technology companies are continuously expanding their market share with technological advantages. In the future, as the market further matures, competition will become more intense. At the same time, with the increasing social attention to environmental protection, digital logistics is also actively exploring sustainable development paths. For example, by optimizing delivery routes and reducing empty mileage, energy consumption and carbon emissions can be reduced. Through a circular logistics system, the generation of packaging waste can be reduced.

In the national "Fourteenth Five-Year Plan and 2035 Vision Outline," logistics and modern logistics are mentioned 19 times in the vision plan. At the policy level, the country has successively issued policies such as the "Construction and Implementation Plan for the Digital Logistics Standard System" and the "Support Measures for Technological Transformation of Logistics Enterprises," which specify specific requirements for digital logistics standardization, intelligence, and networking. At the same time, according to the "2021 Double 11 Green Logistics Report," during the "Double 11" period in 2021, various types of businesses and consumers empowered by green logistics throughout the entire supply chain reduced carbon emissions by 5.3 million tons. Not only Alibaba's Cainiao, but also major domestic express delivery companies have set up recycling bins for express packaging materials in their branches to practice green recycling. Many countries and regions are introducing relevant policies to support the development of digital logistics.

1. **Scenario Analysis**

Digital logistics scenarios and requirements can be decomposed from the following aspects.

① Warehouse Management: In digital logistics scenarios, warehouse management is an important part. By introducing technologies such as the Internet of Things and big data, real-time monitoring, intelligent scheduling, and automated operations of goods in the warehouse can be achieved, improving the operational efficiency and storage capacity of the warehouse. At the same time, digital warehouses can effectively reduce human error rates and improve the accuracy of storage and retrieval.

② Distribution management: In digital logistics scenarios, distribution management is also a key part. By introducing technologies such as artificial intelligence and big data, the distribution routes can be optimized to improve distribution efficiency and reduce transportation costs. At the same time, digital distribution can also achieve functions such as intelligent scheduling, real-time monitoring, and automated tracking, improving the accuracy and timeliness of distribution.

③ Order Management: In the digital logistics scenario, order management is a core component. By introducing technologies such as artificial intelligence and big data, orders can be intelligently processed and automatically tracked, improving the efficiency and accuracy of order processing. At the same time, digital order management can also achieve personalized services, intelligent recommendations, and improve customer satisfaction and loyalty.

④ Customer Service: In the digital logistics scenario, customer service is also a very important component. By introducing technologies such as artificial intelligence and big data, customer needs and feedback can be analyzed and processed in real-time, providing more personalized and professional services. At the same time, digital customer service can also achieve automated replies, intelligent replies, and improve customer satisfaction and loyalty.

⑤ Supply Chain Collaboration: In the digital logistics scenario, supply chain collaboration is also a crucial aspect. By introducing technologies such as the Internet of Things and big data, it is possible to achieve information sharing and collaborative operations among various links in the supply chain, thereby improving the operational efficiency and response speed of the entire supply chain. At the same time, digital supply chain collaboration can also enable functions such as prediction and warning, and intelligent decision-making, thereby enhancing the competitiveness and stability of the entire supply chain.

Taking cold chain logistics as an example, this is an introduction to the overall scenario and solutions.

Cold chain logistics refers to a logistics system in which refrigerated and frozen goods are always kept in a specified low-temperature environment throughout production, storage, transportation, and sale to ensure the quality and performance of the goods. The applicable scope of cold chain logistics includes fresh agricultural products, pharmaceuticals, chemicals, etc., and its characteristics are large investment, high operating costs, and technological complexity. Cold chain logistics has a wide range of application scenarios, including fresh agricultural products, pharmaceuticals, chemicals, the catering industry, and the retail industry.

In the cold chain logistics scenario, the cold storage is one of the key facilities. According to different classification standards, cold storage can be divided into multiple types. For example, based on temperature classification, cold storage can be divided into high-temperature cold storage ($0 \sim 5$ ℃), medium-temperature cold storage ($-5 \sim 0$ ℃), low-temperature cold storage ($-10 \sim 0$ ℃), and ultra-low-temperature cold storage ($-18 \sim 0$ ℃). Based on usage classification, cold storage can be divided into agricultural processing type, food cold processing type, logistics distribution type, pharmaceutical vaccine type, etc.

In the cold chain logistics scenario, the role of cold storage is very important. Firstly, cold storage can ensure the quality and safety of goods. For fresh agricultural products, low temperature can inhibit the growth and reproduction of microorganisms, thereby extending the shelf life of goods; for drugs and chemicals, low temperature can ensure their stability and safety. Secondly, cold storage can provide warehousing services, ensuring the logistics andtransportation of goods. In addition, cold storage can also provide storage services for raw materials and semi-finished products in production. The data that needs to be monitored in the cold storage includes: temperature and humidity in various areas, air conditioning parameters, water accumulation in key areas, cold storage door status and time.

In addition to cold storage, professional refrigerated trucks and refrigerated containers are also needed during transportation. The data that needs to be monitored inside the refrigerated truck includes temperature and humidity inside the truck, air conditioning parameters, groundwater, door opening and closing status and time, and GPS trajectory.

Due to the large deployment range of the collection terminal in the cold storage, moderate collection frequency, and easy adjustment, it is suitable to access the network using LoraWAN or NB-IoT; the collection terminal in the refrigerated truck has a smaller deployment range, but the terminal has higher mobility and requires real-time sampling, so it is suitable to access the network using 4G/5G wireless technology.

The data collected by the terminal is uploaded to the monitoring cloud platform, which needs to achieve the following functions.

① Monitor and integrate important parameters and high-granularity temperature data of the cold chain turnover warehouse.

② Summarize the environmental parameters of all warehouses within the jurisdiction.

③ Record, present, and manage data through a platform.

④ Real-time monitoring of alarms for exceeding the threshold and abnormal alarms based on big data analysis by managing device alarm thresholds.

⑤ Long-term storage of historical data.

Logistics-related personnel can view relevant data from the client, and the client needs to achieve the following functions.

① Multi-level supervisor settings.

② Multiple clients can view temperature conditions.

③ Data sharing and query functions for cold storage, supermarkets, and transportation vehicles.

④ Automatic real-time recording of temperature data, automatic generation of temperature data reports.

⑤ Alarm message notification, alarm message query, voice alarm.

2. Project Scheme Design

The design of the cold chain logistics project needs to consider the following content:

1) Project Objectives

Clearly define the objectives of the project, such as meeting specific market needs, improving cold chain logistics efficiency, reducing costs, etc. Ensure that the objectives are measurable and achievable.

2) Monitoring of Cold Storage and Refrigerated Vehicles

Based on project requirements, design a reasonable deployment of collection terminals, including storage area division, deployment of collection devices and gateways, installation and configuration of network devices, etc.

3) Information Platform Design

Establish a comprehensive cold chain logistics information platform system to achieve real-time updating and processing of information from cold storage and refrigerated trucks. The platform should have functions such as data display and monitoring, historical data viewing, and abnormal data alerts.

Based on the scenario requirements, we propose the following system architecture, as shown in Figure 3.3.8.

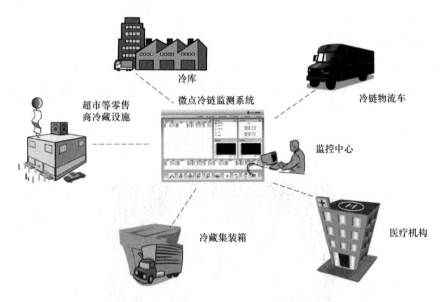

Figure 3.3.8　Architecture Diagram of Cold Chain Logistics System

The detailed design architecture is shown in the following Figure 3.3.9.

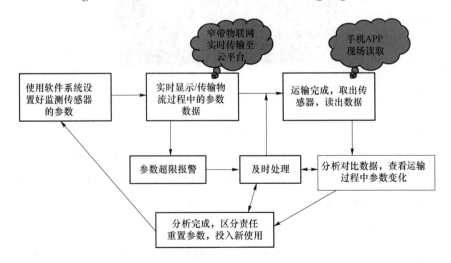

Figure 3.3.9　Detailed Design Architecture Diagram of Cold Chain Logistics System

The following devices are used in this solution.

① Cloud platform.

② Monitoring center.

③ 5G gateway.

④ General-purpose acquisition control module.

⑤ Narrowband IoT collector.

⑥ Digital temperature sensor.

⑦ Digital CO_2 sensor.

⑧ Digital temperature sensor.

Below is a detailed introduction to some commonly used devices.

(1) Narrowband Internet of Things Multi-Channel Collector LTM9103, is shown in Figure 3.3.10.

Figure 3.3.10　LTM9103 Appearance Diagram

① Standard LoRaWAN design, customizableNBIoT/Cat1 communication.

② Can collect 1~20 points of high-precision digital temperature signals.

③ Can collect 2 channels of standard switch signals (connected to access control/water immersion and other sensor signals).

(2) Waterproof Temperature Sensor LTM8877, is shown in Figure 3.3.11.

① Precision digital temperature sensor, supports one-wire transmission of multiple sensor data.

Figure 3.3.11 LTM8877 Appearance Diagram

② Low power consumption, high stability, maintenance-free.

③ Supports remote power supply.

④ Comes with a laser-engraved 64 bit globally unique serial number at the factory.

⑤ Widely used, it has undergone rigorous long-term testing in various field scenarios.

⑥ Digital thermometer measurement range: $-55 \sim 80$ ℃.

a. Temperature resolution: up to 0.062 5 ℃.

b. Temperature measurement accuracy: better than ± 0.5 ℃.

⑦ High-quality stainless steel tube packaging, waterproof, moisture-proof, and rust-proof.

(3) Cloud Platform, is shown in Figure 3.3.12.

① Integrate important parameters of cold chain logistics, including key indicators in various dimensions.

② Can monitor real-time and historical data, provide pre-analysis and alarm reminders.

③ Can monitor and report by zones, and automatically generate integrated charts.

④ Can set warning thresholds and actively send out alarm prompts when data is abnormal.

3. Question and Thinking

Question: The deployment of a logistics monitoring system will undoubtedly increase the investment costs in the cold chain logistics industry. How can cost control be effectively implemented in this scenario?

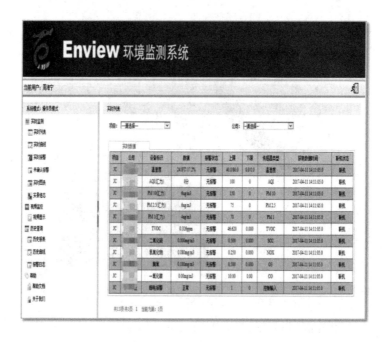

Figure 3.3.12 Cold Chain Logistics Monitoring Platform Data Viewing Page

3.4 商业数字化升级领域——智慧零售

本节案例由北京中兴协力科技有限公司提供,另外,项目创始公司马蹄铁科技也是由北京邮电大学优秀毕业生创建的公司。

3.4.1 行业背景

智慧零售系统是指基于大数据、互联网、物联网、云计算等先进技术,将传统的零售业务与智能技术相结合,实现全程信息化、智能化、自动化的一种新型零售模式。智慧零售系统可以为客户提供更加高效、个性化的购物体验,同时为企业提供更加精细、高效的管理手段,提高企业的运营效率和盈利能力。

2017年3月,全国政协十二届五次会议在人民大会堂举行第二次全体会议,苏宁控股集团董事长张近东,就未来零售发展做了题为《大力推动实体零售向智慧零售转型》的发言,其主要内容是"在零售环节中融入互联网和大数据以及人工智能等先进科技,使得用户、支付和商品等零售要素实现数字化,采购、服务以及销售等运营实现智能化的一种零售新形式",明确提出未来零售就是智慧零售。

智慧零售系统的功能包括但不限于：

① 智能选品：通过分析客户行为、市场需求和竞争情况，智慧零售系统可以自动筛选出适合销售的商品，提高选品的准确性和市场竞争力。

② 智能定价：根据市场需求、商品成本和竞争状况，智慧零售系统可以自动调整商品价格，确保价格合理且具有竞争力。

③ 智能库存管理：通过实时监测商品销售情况，智慧零售系统可以自动调整库存，避免缺货或积压现象，降低库存成本。

④ 智能营销：根据客户行为和偏好，智慧零售系统可以制定个性化的营销策略，提高营销效果和转化率。

⑤ 智能客户服务：通过智能语音交互、在线客服等方式，智慧零售系统可以提供更加便捷、高效的客户服务，提升客户满意度。

要建立完善的智慧零售系统，需要结合多种技术和设备，包括传感器、摄像头、RFID 技术、人工智能等。同时，要建立强大的数据分析和处理能力，对海量数据进行实时分析和处理，挖掘出有价值的信息，为企业决策提供支持。

智慧零售系统的优势主要体现在以下 5 个方面：

① 提高运营效率：智慧零售系统通过自动化和智能化的管理方式，提高了企业的运营效率。例如，通过智能库存管理和智能订单处理等，可以减少人力成本和操作时间。

② 精准营销：通过分析客户行为和偏好，智慧零售系统可以制定更加精准的营销策略，提高营销效果和转化率。这种个性化的服务方式可以更好地满足客户需求，提高客户满意度。

③ 提升客户体验：智慧零售系统可以提供更加便捷、高效的客户服务，例如，智能语音交互、在线客服等。这种智能化的服务方式可以减少客户等待时间，提高服务质量和客户满意度。

④ 增加销售额：通过精准营销和个性化服务，智慧零售系统可以增加销售额和市场份额。同时，智能定价和促销等功能也可以提高销售额和利润率。

⑤ 降低成本：智慧零售系统可以降低企业的运营成本和人力成本。通过自动化和智能化的管理方式，可以减少人力投入和操作时间，从而降低成本。

总之，智慧零售系统的优势在于可以提高运营效率、精准营销、提升客户体验、增加销售额和降低成本等方面。随着技术的不断进步和应用，智慧零售系统将成为未来零售业的主流趋势。

但智慧零售系统也面临一定的局限性，主要表现在以下 5 个方面：

① 技术成本高：智慧零售系统需要结合多种先进的技术和设备，包括传感器、摄像头、RFID 技术、人工智能等。这些技术和设备的成本较高，对于一些小型企业来说可能难以承受。

② 数据安全风险:智慧零售系统涉及大量的客户和商业数据,如何保障数据的安全和隐私是一个重要的问题。企业和客户需要采取有效的安全措施和技术手段,防止数据泄露。

③ 系统集成难度大:智慧零售系统需要与各种设备和系统进行集成和交互,如何实现不同设备和系统的互联互通是一个难题。因此,需要制定统一的标准和规范,方便不同设备和系统之间的集成和交互。

④ 人才短缺:智慧零售系统需要具备相关技术和管理能力的人才进行建设和维护。然而,目前具备相关技能和经验的人才较为短缺,企业需要加强人才培养和引进。

⑤ 法律法规限制:随着智慧零售系统的普及和应用,相关的法律法规和监管政策需要跟进和完善。企业和政府部门需要遵守相关法律法规和政策规定,确保智慧零售系统的合规性和合法性。

3.4.2　项目方案

根据部分时尚服饰零售企业的市场需求,马蹄铁科技推出了一套智慧零售方案,该方案可以实时获取门店单货品的货位、加购、试穿、成交信息,并根据对单货品的流量分析结果来打造完美时尚数字门店。

该方案使用的产品服务架构如图 3.4.1 所示。

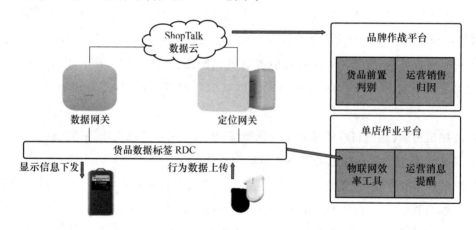

图 3.4.1　智慧零售方案产品服务架构

该方案中涉及 3 个部分:采集终端,云平台,客户端。

使用的采集终端有两类:电子标签和网关。其中,电子标签有两种:电子标签(用于显示货品价格信息)和货品数据标签(用于识别货品行为)。网关也有两种:数据网关(用于下发商品的价格展示信息)和定位网关(用于接收货品数据标签的上传信息)。

终端与网关之间的通信使用的是 AOA(Angle of Arrival)技术,AOA 是一种用于定位的蓝牙技术,它可以通过收集来自多个基站(蓝牙网关)的信号信息,来确定蓝牙设备的准确位置。蓝牙网关主要用于收集周围蓝牙设备的信号信息,包括信号强度、到达时间和角度等。这些信息被发送给后台处理系统进行定位计算,因此,蓝牙网关的安装位置对于定位精度和性能至关重要。AOA 的终端标签尺寸小、功耗低,使用一节纽扣电池便可以运行一年以上,但网关功耗较高,因此一般使用有线电源供电。

通过 AOA 网关和终端部署,系统可以实现以下功能:

① 货品流量监测:实现精细到单货品颗粒度的行为识别,计算拿动次数。

② 货位精确识别:实时洞察货品的出样情况和出样陈列位置,识别模特出样货品。

③ 货品全局改价:实时远程调整价格以及促销信息,减少人员工作量,降低出错率。

④ 货品防损防盗:更高识别精度且更敏捷的防盗方案,实现防盗和货品追溯功能。

云平台在整体方案中有着承上启下的作用,一方面,与采集终端进行消息交互,实时获取商品信息,另一方面,与客户端对接,为门店管理人员提供可视化的数据。云平台的功能主要有 2 个:货品前置管理和销售运营归因。

云平台接收采集终端上传的数据,可以在第一时间获取货品的基本信息(拿起次数、试穿次数、成交次数等),如图 3.4.2 所示。根据这些信息可以判断出货品的销售趋势,如果是潜在畅销款和爆款,便可以通知运营方进行曝光和推广支持,同时提前对供应链进行前置翻单;如果是试穿转化率异常的滞销款和风险款,则可以通知运营方及时折扣出清。

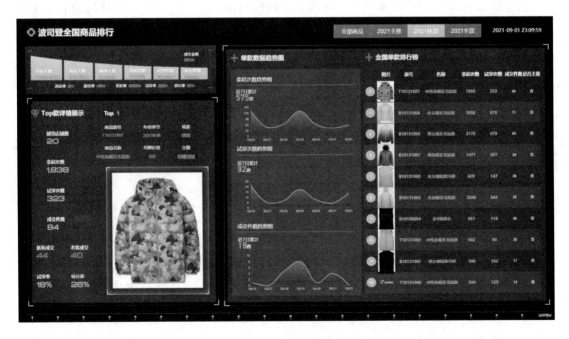

图 3.4.2 智慧零售系统云平台数据查看页面

客户端,即零售终端作业平台,是面向门店管理者的可视化工具。主要功能有:运营消息提醒和终端效率工具。

云平台将整理好的货品数据通过短信或公众号的方式,定期下发通知,以便门店工作人员更直观、高效地看到各个货品的转换率和销售信息,并根据平台下发的后续销售建议及时对货品摆放进行调整。

3.4.3 合作案例

目前,马蹄铁科技推出的智慧零售方案已与多个时尚品牌展开深度合作。

1. 波司登

客户简介:专注羽绒服49年,是中国最大、生产设备最为先进的羽绒服生产商。

服务价值:通过用户行为数据抓取,精准且实时掌握商品动态。决策前了解商品在各个区域的表现,及时准确地调整商品铺货策略,从而做出正确的货品翻单。

2. 唯品会

客户简介:主营业务为互联网在线销售品牌折扣商品,涵盖名品服饰鞋包、美妆、母婴、居家等各大品类。

服务价值:针对唯品会线上线下一体化商品,进行货品品类配置,并对场内陈列热区进行分析,助能业务部门实现精细化门店管理,以及单店经营分析。通过转化率分析唯品会场内货品结构,为单店订货、品牌配置、品类配置提供过程数据参考。

3. KC皮草集团

客户简介:KC皮草是全国领先的皮草零售龙头企业,年营收10多亿。

服务价值:马蹄铁科技全面为KC皮草货品上线了智能吊牌设备,实现了全维度触摸、试穿、定位数据的收集。将用户行为数据快速作用于供应链,成功预测出当年潜在热销款和热销色。供应链联动实现了前置反应,同时提升爆品满足率。

3.5 创新创业实践应用案例

本节案例由北京邮电大学国际学院物联网工程专业提供。

随着物联网行业的兴起和发展,各大高校在相关方向的培养和建设也在同步进行。北京邮电大学自2009年以来大力发展物联网技术的研究、应用和转化,并取得了显著的成果。

3.5.1 案例1:基于多源融合感知的智能博物馆消防应急系统

2015年,云南省大理古城发生火灾,过火面积约300 m^2,600多年的历史古迹全被烧毁,尽管大火已经扑灭,财产损失可以统计,但文化的损失不可估量,曾经的拱辰楼在火灾过后只剩下一个框架。党的十八大以来,以习近平同志为核心的党中央高度重视中华文化的传承发展和文化遗产保护工作。习近平总书记多次参观博物馆和纪念馆,并对文化遗产保护工作做出批示和指示。"大火过后,刺心的痛在这座小城蔓延",守护文化,守护博物馆等古建筑群体是该项目创意的初衷。

该项目的研究意义在于响应国家对文化遗产保护的重要号召,设计一套面对博物馆的多源融合感知智能消防应急系统,解决博物馆发生火灾这一复杂场景下消防出警困难的问题。同时,借鉴国内外的研究成果和经验,结合信息物理系统的发展,构建智慧消防试验平台,并提出实时态势感知监控和数据获取,以及消防灭火给水等方面的技术挑战和未来发展方向。这将为文物火灾保护工作和智慧城市建设提供重要的技术支持和实践平台。

1. 项目研究的主要内容

这个项目主要研究的是面向博物馆的多源融合感知智能消防应急系统,旨在解决博物馆发生火灾这一复杂场景下消防出警困难的问题。系统通过基于视觉/位置感知的消防应急通道状态监测系统,能实时监控消防通道的状况。在救援和疏散方面,应急救援的最优路径导航系统可以根据火情信息实时规划最快安全的救援及疏散路线。同时,基于自注意力机制模型进行步伐判断的计步器可以对用户的疏散步伐进行实时监测和反馈,提供实时人员疏散信息。此外,在日常巡检和器材方面,该项目的消防器材态势信息管理系统能提供数字化管理的消防器材,并对指定器材进行查询。这4个模块的综合应用,可以提升博物馆的火灾应急响应能力和消防安全管理效率。

2. 项目分析

(1) 痛点分析

该项目主要针对我国多数的文物保护单位,尤其是对历史文化传承中较为完整的古建筑群进行消防安全管理的优化。这些文物保护单位多面临着消防安全管理的困难,主要问题包括:建筑结构复杂,火灾隐患频繁,消防报警系统不完善,以及在火灾发生时无法快速有效地规划疏散路径等。其中,楼宇的复杂性和防火分区的间距小增加了火灾的风险,使得消防人员在火灾发生时难以迅速抵达并找到最佳灭火路径。对此,第一,该项目专注于室内导航,以解决消防员在室内难以快速准确地到达指定地点的问题。第二,在火灾发生过程中，人群或者物品倒塌将导致消防通道堵塞，该项目可通过传感器进行堵塞识别，避免在疏散和救援过程中遇到此问题。由于消防报警系统不健全，故各种隐患往往难以被及时发现和排除。再者，建筑复杂的内部结构使得规划疏散路径成为一个严峻的挑战，在火灾发生时，这可能会阻碍人员的及时、安全撤离。此外，楼宇复杂度高将导致消防器材日常管理困难。为此，该项目对消防器材管理进行了数字化管理，对器材进行定期更新和类型管理，并通过数字化监督和公告发布，有效保证消防器材的及时更换和正确使用，从而提高应急救援的效率和效果。

（2）需求分析

本项目的目标用户是需要加强火灾应急系统的文物保护单位。这些单位通常承担着保护我国重要历史文化遗产的责任，其中包括古建筑群。为了更好地完成这一任务，这些单位需要一套强大且易于操作的消防系统。这个系统应当具备日常的消防系统监控，能够自动检测和上报火灾隐患，精确定位复杂楼宇的室内空间，以及在火灾发生时能够即时规划出警及疏散路线等功能。运用可以满足这些需求的系统不仅能大大降低火灾发生后的损失，提高人员的安全，也能响应国家应急管理局对防范火灾等紧急情况的指示，体现国家对古建筑群保护的重视。因此，该项目的系统具有很大的实用意义，未来应用范围也将十分广泛。

3. 系统概要设计

发现、通知：消防险情发生时，系统通过多种传感器自动感知火情信息并立马将其通报给消防队和管理处，并依据决策进行下一步行动，确保了消息的即时性和可靠性。

出警、撤离导航：在救援过程中，系统会根据火情信息，帮助消防人员规避风险房间，规划一条快速安全的路径，将消防人员导航到火灾发生房间，同时为游客规划出一条安全疏散路线，并且火势的发展也会进一步反馈回导航，更新路网。保证了游客的安全，并提高了救援效率和准确性。

现场救援：灭火过程中，消防人员可以通过系统查询附近消防器材，也可依据需求查询指定消防器材位置。此外，系统可以依据探头进行目标检测，实例分割，了解火灾发生后室内的人员疏散情况以及各个通道的畅通情况，辅助消防救援。

日常维护：日常维护方面，系统设有消防器材态势感知信息管理模块，可对消防器材信息进行编辑修改，支持多维度器材查询显示器材位置等详细信息，可依据设置对消防器材进行损耗预警，并可依据算法对火灾风险进行静态评估，打分预警。此外，系统可针对计算机视觉捕捉到的画面分割图像，识别消防通道是否堵塞，若发现问题则及时报警。保障了日常维护工作的顺利进行。

智能消防应急系统架构如图3.5.1所示。

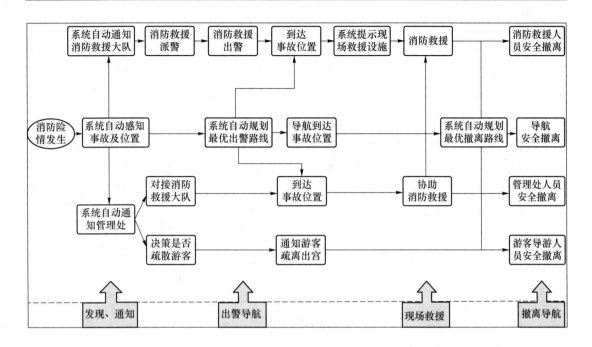

图 3.5.1 智能消防应急系统架构

4. 模块的设计与实现

基于视觉/位置感知的消防应急通道状态监测系统设计与实现：对计算机视觉捕捉到画面进行图像分割、目标检测，查看消防通道是否堵塞，以及检测各通道人员数量和疏散情况。

面向应急救援的最优路径导航系统设计与实现：火灾发生时，依据火情信息，为消防人员筛选出最快的到达着火点的安全路线，为建筑内的人员导航一条快速、安全且不冲突的撤离火灾现场路线，并将火情的发展进一步反馈回导航，更新路网。

面向应急救援的消防器材态势信息管理系统设计与实现：实现消防器材维护和管理的数字化，支持多维度查询管理消防器材，具有器材损耗预警、位置提示和火灾风险静态评估功能，火灾时可实现火灾态势感知和预测。

面向应急救援的基于自注意力机制模型进行步伐判断的计步器设计与实现：步伐判断系统主要用于实时监测用户的步伐，并将步伐变化反馈回计步器。在性能方面，步伐检测的准确性和实时性是关键的指标。

5. 项目研究成果

多源融合定位技术方案：针对博物馆人员智慧管理时空信息缺失难题，面向文物保护的多源融合定位系统总体技术方案如图所示。系统包括一个多源融合定位云平台、四种智能感知终端及五类位置信息。多源融合定位云平台结合位置信息，对接博物馆现有 OA 系统，提升博物馆现有系统智能化水平。四种智能终端包括智能卡、手机、游客小程序和消防人员专用终端。智能卡终端基于 RFID、UWB 的位置服务，将用户位置信息通过基站上传到多源融合定位云平台对接博物馆 OA 系统,管理人员可以在 A 终端查看用户位置、进行时空信息统计等。手机终端主要面向巡检人员等,提供室内外无缝定位、位置查看等扩展服务。同时,手机终端可与智能卡自动绑定,在用户携带智能卡时进行多设备系统定位,进一步提高系统定位精度。游客小程序终端通过微信小程序接入云平台,业主可以通过后台查看人流分布情况,合理导流。消防人员专用终端是在巡检人员功能的基础上集成了 MESH 自组网设备,具备在火灾通信中断情况下依旧坚韧导航的通信能力。五类位置信源包括读卡器、UWE 基站、蓝牙信标、Wi-Fi 信标及 MEMS 器件。读卡器可通过智能卡提供关键点定位信息、用户 ID 信息及身份验证功能。UWB 基站可提供高精度位置估计。由于 UWB 基站的部署成本昂贵,故仅在部分高精度定位区域进行安装,其他区域采用低成本的蓝牙、Wi-Fi 信标方案来提供全空间位置信息。最后,MEMS 器件提供相对位置变化信息,优化系统定位的连续性。智能应急消防系统定位功能架构如图 3.5.2 所示。

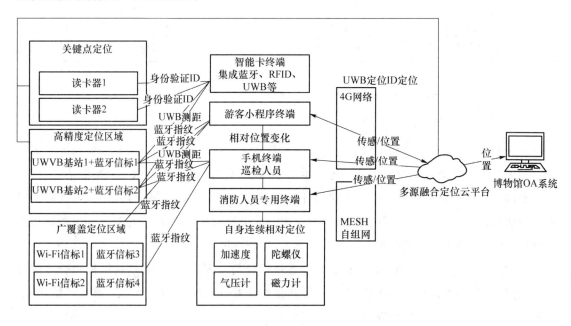

图 3.5.2 智能应急消防系统定位功能架构

基于视觉/位置感知的消防应急通道状态监测系统:在前端开发技术方面,该项目使用了流行的 HTML、CSS、JavaScript 和 layUI 等技术,以实现用户友好的界面和良好的交互体验。后端开发技术方面,该项目选择了一些流行的 Web 后端开发技术框架,如 Maven,用于处理后端的请求和提供 API 供前端调用。数据库设计方面,该项目根据需求设计了合适的数据库表结构,并建立了索引以提高查询性能。该项目使用了一些常见的数据库管理工具,如 MySQL 和 Navicat,以优化数据库连接的效率和性能。此外,该项目还应用了数据分析技术,包括一些常用的数据统计算法,以便对消防器材进行图表反馈和数据可视化,为消防工作提供有益的数据支持。

面向应急救援的最优路径导航系统:该模块的技术框架基于计算机视觉、深度学习(特别是 YOLOv5 算法)和数据处理等关键技术。计算机视觉技术被应用于获取和处理消防应急通道的图像数据,通过图像采集设备获取实时场景图像,并利用图像处理算法进行预处理和增强,以提高图像质量和准确性。深度学习算法 YOLOv5 则发挥关键作用,它能够实现高效的目标检测和识别,通过训练模型来识别图像中的人物、物品和状态,例如,判断消防通道是否被堵塞或存在障碍物。数据处理方面,该系统对采集到的图像数据进行清洗、分析和可视化。

面向应急救援的消防器材态势信息管理系统:为了实现面向应急救援的最优路径导航系统,该项目主要采用了 A 算法。A 算法作为一种启发式搜索算法,被用于在图形结构中寻找最短路径或最佳路径。它通过使用启发式函数评估节点的优先级,并使用代价估计和节点扩展的方式来逐步搜索目标节点。它在路径规划、导航和规划领域广泛应用,能够高效地找到最优解决方案。

面向应急救援的基于自注意力机制模型进行步伐判断的计步器:在步伐检测方面,该项目使用了 Python 作为自注意力机制神经网络模型的开发语言,并借助 NumPy 和 Matplotlib 进行高效率的文件读取、数据预处理,以及训练过程模型损失变化的可视化绘制。当然,该项目也借助了使用 Android SDK 开发的步伐数据采集软件进行数据收集。在开发的过程中,该项目采用了版本控制来辅助开发,通过及时的版本回滚,该项目可以规避由重大操作失误或者开发错误而导致的程序无法运行的问题。同时,该项目会在采集到的数据中随机抽取出一部分作为测试集,并将其输入至解码器计算模型检查模型的准确度,确保模型运行在较高的准确度上。

6. 成果测试

消防管理网页:如图 3.5.3 所示,消防器材管理系统实现了对消防器材和巡检人员的数字化监督和信息化管理,其成果包括:器材管理,可以查询、修改、增删实体消防器材的详细信息和位置;巡检人员管理,可以查看、修改、更新巡检人员的详细信息和巡检范围。此外,还可以根据消防器材的数据进行数据分析和预测,实现消防态势感知等功能。在性能方面,如果系统设计合理,各个组件间配合良好,系统的运行效率和稳定性应该会较高,能够满足日常消防工作的需求。

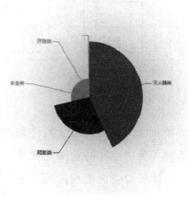

图 3.5.3 智能应急消防管理页面

目标检测:如图 3.5.4 所示,通过对大量数据进行端到端的训练,实现将单阶段目标检测模型分解为特征提取和目标检测两个模块。特征提取模块负责从输入图像中提取特征,目标检测模块则利用提取的特征对目标位置和类别进行预测。目前,目标检测可以在完成输入图片或视频后,通过算法输出分析好的图片或视频,对输入的图片或视频中的人、物进行分割并标注。

图 3.5.4　图片/视频目标检测效果图

面向应急救援的消防器材态势信息管理系统:为了实现导航路径,该项目引入了 A 星算法来处理路线规划的需求。作为一种启发式搜索算法,A 星算法能够在图形数据表示的环境中找到最优的路径。目前,基于 A 星算法,该项目实现了在平面图上的动态路径生成,即在给定的数据地图上,初步建模一个障碍物房间,然后在两点之间实现障碍物规避生成最短的路线,如图 3.5.5 所示。

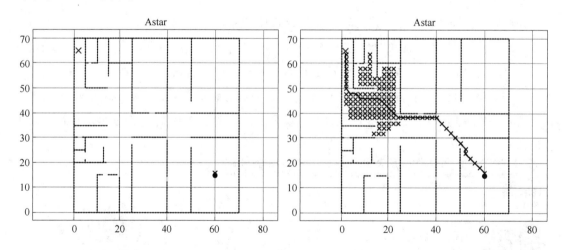

图 3.5.5　搜索算法效果图

面向应急救援的基于自注意力机制模型进行步伐判断的计步器:该项目的步伐检测模型可以极其准确地实时检测出使用者是否正在行走,并且可以规避其他计步器实现中因原地来回晃动计步设备而造成的假计步以及因放置于不同躯体位置导致的计步不准确的问题。首先,该项目使用 Python 从 csv 文件中读取数据,并以数组的形式存储在变量之中;然后,利用

numpy 和 pytorch 对将用于训练的数据矩阵进行转置、掩码，对标签进行编码，将文本转换为对应的数字；最后，调用 cuda 函数将数据和标签转换为可供 GPU 处理的数据。数据读取完成后，该项目会将数据转换为可供神经网络模型使用的数据集，然后构建一个由位置编码器、编码器和解码器组成的 Transformer 模型，将数据和标签输入神经网络模型进行训练。训练过程中，该项目使用交叉熵损失进行迭代，Adam 优化器进行优化，Matplotlib 绘制训练过程损失与训练步数的折线图，如图 3.5.6 所示。

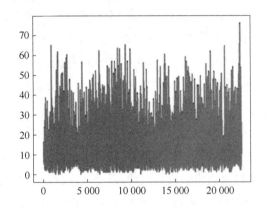 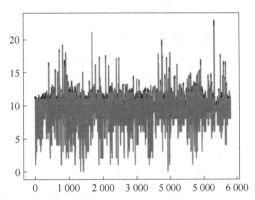

图 3.5.6　训练过程损失与训练步数折线图

3.5.2　案例 2：智能会展系统

1. 项目的背景以及意义

该项目致力于利用"互联网＋"对会展的驱动力，推进传统会展再造，充分利用新一代信息技术，以平台为基础，以低成本硬件为载体，以市场需求为导向，打造线上线下相结合的智能会展系统。该系统将收集、分析、整合会展行业的各类信息，实现对会展多方参会者在公共安全、配套服务、相关活动等方面需求的智能响应，进而全面提升会展行业的运营效率，降低成本，加强项目监管，提高服务水平。

2. 项目研究的主要内容

该项目研发设计的智能会展系统主要包括智能投票系统和人流密度检测系统。智能会展系统巧妙利用软硬件设施，以"互联网＋"为核心，在"互联网＋会展组织""互联网＋展馆管理""互联网＋会展运营""互联网＋会展形式"四个层面进行优化。用户端将采用智能手环的形式。智能手环结合工艺设计专业优势，创新硬件载体的交互方式及外观设计，且手环外观融入了非物质文化遗产、设计师个人创作等。微信小程序作为后端平台上传、收集、分析、整合各类信息，从而便于参会者，特别是会展管理者活动。

整套系统主要分为硬件、云端传输、后端算法及框架以及前端界面。硬件包含 NFC 模块及蓝牙定位模块。智能手环在整套系统中的角色是信息的发出者，智能手环中包括 NFC 射频芯片以及蓝牙射频芯片两个电子标签,分别用于 NFC 投票以及蓝牙定位。对于 NFC 开发端,该项目主要使用的 NFC ANDROID。对于蓝牙模块,该项目选用在室内定位具有较高精度的蓝牙模块进行测距。在开发芯片上,该项目选用 nRF5340 SoC,该芯片支持广泛的无线协议,低功耗蓝牙,具有测向中所有进行到达角度(AoA)和离开角度(AoD)角色功能,以及包括 LE 音频、长距离、2 Mbit/s 高吞吐量和广告扩展功能。前端利用小程序实时公开、查看投票情况,并利用小程序独特通信安全证书进行加密处理,此处返回包初定为 JSON 格式,返回内容体量小,利用 GET 请求即可。前端需要实时更新数据。利用 setInterval 和 JS 的 Promise 来进行异步循环请求,以检测人流密度和人员是否发生变化,并当发生变化时返回新的人流密度和人员信息。后端采用 python+redis,人流信息的特点为数据量大,更新频次快,要求更新速度快等。此外,后端需要高频次处理前端发送的请求,来判断人员是否发生变化并实时更新信息,这要求后端语言处理信息速度快。

该项目充分运用"互联网+"思想,致力于解决参会者与展会者难以获得较好的交流反馈,难以系统地收集统计各个小组的流量以及整体的参会状况问题,利用相关技术促进互联网展会者与参会者的交流,并结合现今可穿戴式设备持有量持续增长的现状,以手环形式提供解决方案。技术方面,蓝牙测距规避了视觉技术无法判定距离,GPS 技术在室内定位不精准的问题,充分保证了产品的综合适应性。用于投票的 NFC 技术以硬件为载体,凭借其低成本、便携轻便、低延迟、交互性好的特点实现了快速且交互体验好的投票方式。与现有的微信扫码投票方式相比,其具有网络要求宽松,数据可缓存在安卓应用本地,在合适时机可上传网络;用户使用方便,无须频繁操作;主办方收集查看数据直观便捷等优势。文创方面,不同活动通过独有的手环样式和后期发放的手环饰品可提升此次活动的文化价值,通过发行限量样式等形式可调动参会者的积极性,提升活动的知名度,同时也为参会者留下了具有实用意义的纪念产品。

该项目的成品应用不仅限于大创展等相关展会,还可以用于人流监控和项目投票统计中,如游乐场、商场等商业场所均有实用价值。该项目可与展会主办方协商,定制手环样式,并在会场部署相关设备,为参会者提供安卓手机端的产品管理应用程序,用于发放手环。此外,后期与展会主办方协商发放定制饰品,实现商用价值。

产品逻辑如图 3.5.7 所示。

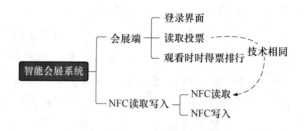

图 3.5.7 智能会展系统产品逻辑

NFC 界面如图 3.5.8~图 3.5.11 所示。

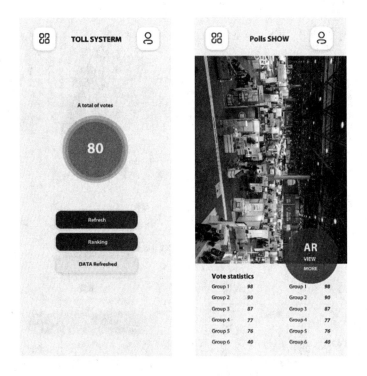

图 3.5.8　投票结果和投标统计界面

图 3.5.9　NFC 读写界面

图 3.5.10 小程序登录界面

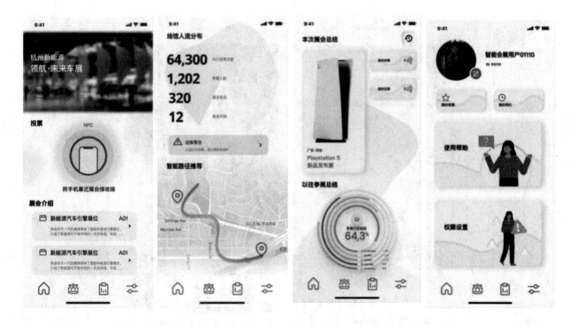

图 3.5.11 小程序显示界面

3. 项目研究的特色

(1) 硬件及蓝牙定位创新

硬件包含 NFC 模块及蓝牙定位模块。智能手环在整套系统中的角色是信息的发出者,智能手环中包括两个电子标签,分别为 NFC 射频芯片以及蓝牙射频芯片,用于 NFC 投票以及蓝牙定位。对于 NFC 开发端,该项目主要使用了 NFC ANDROID。对于蓝牙模块,该项目选用在室内定位具有较高精度的蓝牙模块进行测距。在开发芯片上,该项目选用了 nRF5340 SoC,其可以适配更多的场馆与场地,避免了无线局域网和蜂窝网有时因为在地下而信号不好的情况。低功耗的蓝牙可以降低会场的电费以及其他物业项目费用,有效降低会场的经费支出,节约成本。同时,大多数展会内的人流量较为密集,蓝牙模块可以更好地适应较为拥挤、狭小的区域。

(2) 外观及前端创新

APP 采用扁平化插画风格,并以蓝白为主基调,不仅可以给用户营造更为商业的氛围,提高对展商的关注度,还可以增加页面视觉丰富度。同时,在开发上,多采用卡片组件,创新多种交互形式,提高 APP 的信息展示效率,增加用户和 APP 之间的交互乐趣。页面切换中采用了更平滑的智能效果,除了直接跳转,页面还增加了卡片的平滑移动。以往参展总结部分增加了多张带状条型统计图表,可以最大程度地利用界面,展示出最多的信息。

(3) 后端及数据收集创新

大多数前端利用小程序实时公开、查看投票情况,利用小程序独特通信安全证书进行加密处理,此处返回包初定为 JSON 格式,返回内容体量小,利用的是 GET 请求。前端需要实时更新数据。利用 setInterval 和 JS 的 Promise 来进行异步循环请求,以检测人流密度和人员是否发生变化,并当发生变化时返回新的人流密度和人员信息。由于人流数据需要实时更新,setInterval 算法能够满足实时的请求的功能。后端采用 python+redis,人流信息具有的特点为数据量大,更新频次快,要求更新速度快等。此外,后端需要高频次处理前端发送的请求,来判断人员是否发生变化并实时更新信息,这要求后端语言处理信息速度快。前端部分确定了基本的开发时间线,后端部分了解了核心路径推荐算法。智能路径推荐算法采用 Dijkstra 算法,该算法主要特点是以起始点为中心向外层层扩展,直到扩展到终点。其中,通过比较人流量和距离的远近加权算出各个点位的时间,从而进行推荐。

3.5.3 案例 3:基于物联网的智慧停车装置和线上平台

1. 项目的背景以及意义

随着许多城市的人口和车辆的增加,人们面临的"停车难"问题越来越严重,而大量的车位却没有得到充分利用。本项目旨在建立一套面向用户和城市管理部门的,适用于停车资源紧张地区的停车管理系统。该项目的目标是为有停车需求的用户提供停车位预测和预约分配功能,帮助用户更好地规划行程,提高用户停车效率,缓解停车难问题;为城市管理部门提供智能定价和可视化数据分析功能,提高停车管理水平。

该项目采用 Web 端和安卓用户端相互补的产品模式,为政府决策部门、交管部门、私人企业以及个体车主四大目标服务对象提供技术支持。

该项目在构建边缘设施物联网的基础上,通过对各项停车数据进行汇集和分类处理,将数据直观清晰地展现在可视化界面上,为管理者和决策者提供更高质量的交通服务,从而实现更加精细的运营管理。城市管理部门可以根据系统中停车位占用情况和停车高峰的预测结果做出反应,提前在可能出现拥堵的地段加派警力,以疏通交通堵塞,避免事故的发生,从而缓解车难、管理难的问题。这不仅可以提高人民群众的获得感、幸福感,也对完善民生工程具有重要意义。

2. 项目研究的主要内容

智能停车系统架构如图 3.5.12 所示。

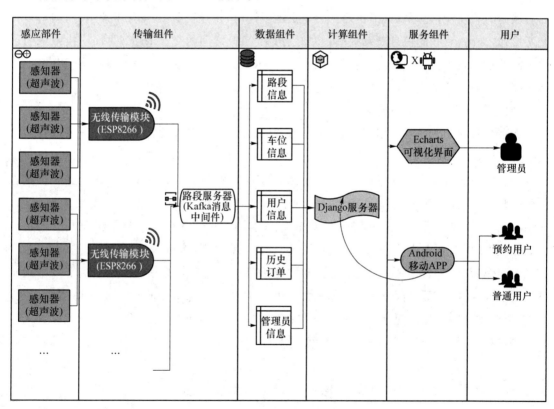

图 3.5.12 智能停车系统架构

该项目完成了车位分配和预测部分的数学建模。通过实验验证,由分配算法来分配车位比用户自行寻找车位,在搜寻时间和从停车位步行到目的地步行时间之和最少降低了 33%。车位分配算法如图 3.5.13 所示。

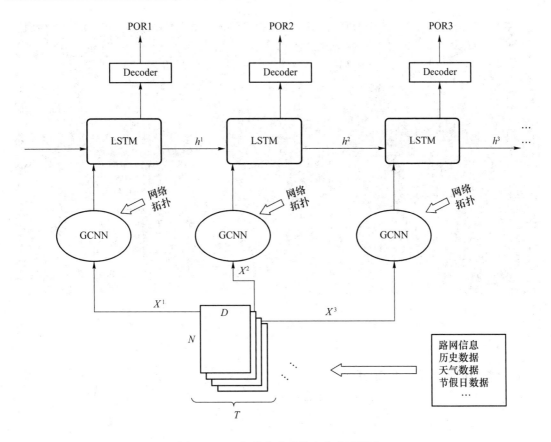

图 3.5.13 智能停车系统车位分配算法

该项目使用 Django 搭建后台,实现了停车分配算法,完成了停车数据库的搭建。数据库中的数据使用了部分深圳市公开的停车数据,用于功能测试。在本地运行程序时,安卓端和 Web 端根据提供的 IP 地址可以成功对后端数据库进行操作。

Web 管理端模块主要使用 Vue 框架进行页面构建。在页面大量的数据可视化模块中,利用 ECharts 将数据进行组织,并转换为直观的饼图、折线图等统计图输出到页面中。通过系统大屏可以清晰、直观地将交通状况、实时预测,基础设施建设等方面的数据进行可视化展示。特别地,它主要反映的是在一定时间范围内路边车位的占用情况。在总体情况界面,管理者可以通过泊位状态得到不同街道处停车位的占用率信息,也可以直观地看到区块利用率,总体占用情况信息。此外,管理者也可以通过实施占用排行、平均占用率、区块占用率、区块 BCU 指数图像得到想要的数据。如图 3.5.14 所示。

安卓端的开发目前主要面向智慧停车引导的用户。开发过程中,功能使用 Android Studio 进行开发,并利用 Gradle 工具进行打包和签名。此外,APP 引入了百度地图 API,并使用 XML 文件储存图形化界面的布局、容器组件等。通过以上工具可以让用户快速访问自己所需功能,同时也增强了安卓客户端的可拓展性、易维护性。

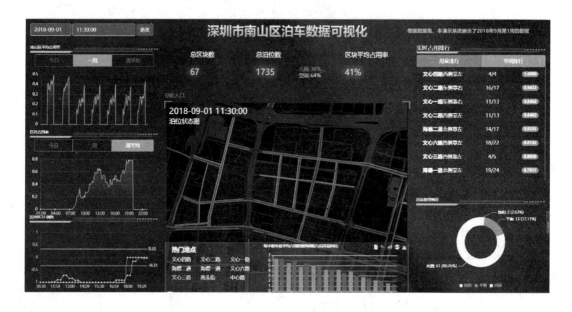

图 3.5.14 智能停车系统 Web 管理页面

1）用户使用流程

用户使用流程如图 3.5.15 所示。

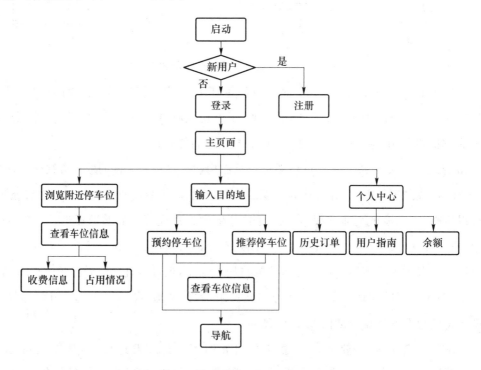

图 3.5.15 智能停车系统用户使用流程

2）功能简析

① 预约车位：用户选择预约服务并输入预约到达时间和目的地。云端服务收到预约请求后利用算法计算目的地附近符合用户需求的停车位，并分配停车位、返回停车位和导航路径。

② 推荐：用户选择停车位推荐服务并输入目的地，云端服务收到目的地停车位推荐请求后利用算法计算目的地附近符合用户需求的停车位并返回停车位和导航路径。

③ 查询：用户点击头像进入个人页面，点击 My appointment 按钮查看用户的预约记录。

④ 查询余额：用户点击头像进入个人页面，点击钱包按钮查看余额。

3）页面架构

安卓客户端的页面架构如图 3.5.16 所示，页面显示图如图 3.5.17 所示。

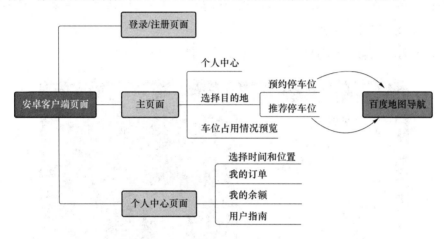

图 3.5.16　智能停车系统安卓客户端页面结构

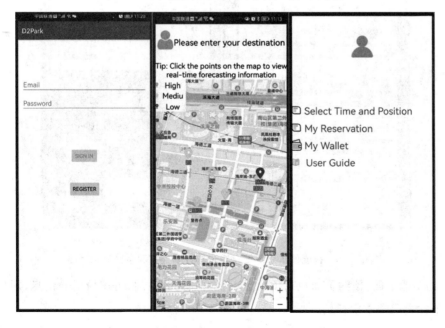

图 3.5.17　智能停车系统安卓客户端页面显示图

(1) 一级页面架构

① 登录/注册页面:在登录/注册页面中,未注册的用户需要点击REGISTER按钮进行注册,已注册用户需要输入邮箱和密码进行登录。

② 主页面:在主页面中,用户可点击地图上的位置点选择行程终点,也可移动地图选择可视范围。

③ 个人中心页面:在个人中心页面中,包括四个部分,分别为选择时间和位置、我的订单、我的余额和用户指南。

(2) 二级页面架构

① 我的订单页面:用户可在订单页面查看所有历史订单,包括订单编号、订单时间、停泊位置等信息,如图3.5.18所示。

② 我的余额页面:用户可进入余额页面查看所剩金额。

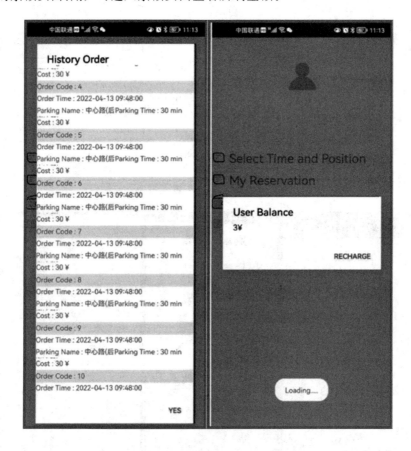

图3.5.18　智能停车系统安卓客户端订单查询和余额显示页面

③ 预约功能页面:预约页面包括预约地点、价格、到达时间、占用情况等信息,用户可选择直接导航前往或取消本次预约。

④ 推荐功能页面:推荐功能页面可显示出为多个用户推荐的停车地点,用户可选择其一

并导航前往,也可选择取消本次推荐。

智慧停车装置产品原型是基于 Arduino 设计,搭配 TowerPro MG90S 金属齿轮舵机、Risym ESP8266 串口 Wi-Fi 模块和 HC-SR04 超声波模块,实现了产品设计的预期功能。

基于 Risym ESP8266 串口 Wi-Fi 模块控制,保证了通信的稳定性并且验证了方案的可行性。HC-SR04 超声波模块进行测距,掌控距离进而检测车辆的驶入驶出。最终控制 TowerPro MG90S 金属齿轮舵机,进行模拟车位锁抬杆放杆。

在初代产品模型上,该项目已经初步实现了预期的产品功能,并经过验证表明,该产品与预期的产品较为相近,且与同类产品相比,该产品具有操控性更高、研发和设备成本更低等特点。

基于 MQTT 协议以及 Kafka 分布式流媒体平台中间件构建系统的消息队列模块。通过引入 Kafka 中间件构建分布式发布订阅系统,实现在高吞吐量情况下信息的稳定获取。构建的分布式的 Kafka 集群大大提高了系统集群的稳定性与效率。通过 Kafka,消费者和生产者稳定传输着由硬件传感器传输的消息信号,是硬件端和后端数据库之间的"桥梁"。

3. 项目研究的特色

该项目的应用可以对停车位的使用情况进行短期预测和长期预测。使用基于注意力机制的时空图卷积网络预测模型对获取到的停车数据、POI 数据、天气数据、路网数据进行整合分析,可实现停车位预测功能。这样既可以帮助用户更快速地判断与决策停车地点,更合理地规划自己的行程路线,又可以帮助管理层对停车位进行合理定价,确保收益。

该项目的整体模型为基于竞争和合作的分配模型。

在预约模型中,存在三种用户:外部用户(不使用平台的用户)、预约用户和普通用户。其中,预约用户和普通用户之间区别在于是否进行预约,普通用户的分配使的是基于竞争关系的推荐算法,而预约用户的分配使用的是基于合作关系的预约算法。

(1) 基于竞争关系的推荐算法

对于不选择预约服务的普通用户,平台会推荐一系列的推荐停车场,并且将停车场连起来形成一条寻找停车位的最佳路径。这条最优路径在设计时要求尽量短,并且沿着该路径能以较高的概率找到停车位,同时保证用户下车后步行到目的地的距离较短。如果未能成功停车,则分配停车路径上的下一个停车位,直到成功停车,或因无下一个停车位而提示推荐失败。

(2) 基于合作关系的预约算法

对于选择了预约服务的预约用户,首先进入等待队列中等待停车位的分配。每次分配时,平台会考虑驾驶距离和步行距离,同时对多用户进行分配,找到一个全局最优的解。分配成功的用户将进入预约队列,分配失败的用户则将返回等待队列等待重新分配。平台会对预约队列中的用户进行周期性迭代优化,分配给该队列内用户停车位的位置会随着距离目的地的距离变近而优化。

考虑多样化需求的停车引导分为基于竞争关系的推荐和基于合作关系的预约两种。基于竞争关系的推荐服务适用于普通用户，平台将通过该分配提供给用户一条寻找车位的最佳路径，帮助用户以较大的概率寻找到离目的地距离近的停车场，该种分配提供给用户最大可能找到停车位的停车引导服务。基于合作关系的预约服务则是在保证一定的公平性的前提下，通过对全体用户进行统一分配，找到全局最优的解决方案，提供有保障的停车预约服务。该项目考虑了外部用户对预约用户和普通用户的影响，优化了两类用户的停车体验，同时减少了两类用户对外部用户的负面影响，以实现停车位分配的公平、高效、可扩展。

参 考 文 献

[1] 刘云浩.物联网导论[M].4版.北京:科学出版社,2022.
[2] 莱亚.物联网架构设计实战:从云端到传感器[M].陈凯,译.北京:清华大学出版社,2021.
[3] 孙昕炜,李江,王恒心.工业物联网应用与实践[M].北京:清华大学出版社,2022.
[4] 高泽华,孙文生.物联网:体系结构、协议标准与无线通信:RFID、NFC、LoRa、NB-IOT、WiFi、ZigBee与Bluetooth[M].北京:清华大学出版社,2020.
[5] 甘泉.LoRa物联网通信技术[M].北京:清华大学出版社,2021.

附录　LoRaWAN 实验箱说明

本书中涉及的 LoRaWAN 相关实验是基于微点科技的实验箱进行的,除了书中提到的设备,该实验箱还包含了多种测量终端,能满足多种 LoRa 场景下的网络部署和数据采集。

LoRaWAN 实验网络架构如图 1 所示,终端 End Nodes 使用 LoRaWAN 实验箱内设备,网关 Gateway 使用 RHF2S024,网络服务器 Network Server 使用开源 NS 服务器,应用服务器 Application Server 使用优物联云。其中,终端设备实验箱外观如图 2 所示,内部布局如图 3 所示。

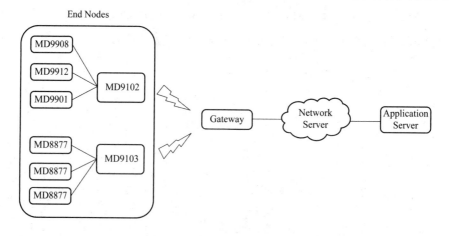

图 1　LoRaWAN 实验网络架构

图 2　LoRaWAN 实验箱外观

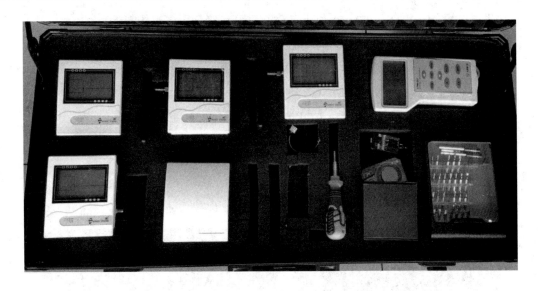

图 3　LoRaWAN 实验箱内部布局

实验箱中包含设备如表 1 所示。

表 1　实验箱设备一览表

设备型号	设备名称	设备功能
MD9102	ITU 数字总线采集转换模块	用于挂接多个传感器,并将多个终端数据一次上传
MD9103	多通道数字温度传感器采集模块	用于挂接多个数字温度传感器,并将多个终端数据一次上传
MD9908	TVOC 传感器	用于采集现场总挥发性有机化合物浓度
MD9912	AQI 粉尘传感器	用于采集现场 PM1、PM2.5、PM10、空气质量数据
LTM9901	微型温湿度传感器	用于采集现场环境温湿度数据
MD8877	数字温度传感器	用于采集现场多点温湿度数据

下面对各个设备的功能进行详细的说明与介绍。

1. MD9102

MD9102 是遵循 LoRaWAN 协议规范的无线通信 ITU 总线产品采信转发模块,可以配合各种 ITU 总线设备,把现场温度、湿度、TVOC、空气粉尘浓度、AQI 指数及其他标准模拟量(如电流电压电量传感器、风向风速传感器等)、开关量信号(如烟雾报警、水浸报警)多种数据上传到 LoRa 云平台。配合不同类型的终端可以灵活组建多种大小不同的现场监测应用系统,实现物联网的测点分散、汇聚平台、自由访问。使用 MD9102 搭建监测系统可以达到一路 LoRa 无线信号,多种数据汇总上传的无线传输方式。

技术指标如下所示:

① 网络接口:默认 LoRa C 模式。

② 工作频段:470~510 MHz。

③ 供电电源：+5V DC,200 mA 以上(电源双接口,符合更多现场需求)。

④ 外形尺寸：98 mm×85 mm×45 mm。

⑤ 最多可支持 32 路温湿度测量(配合 MD9901)。

⑥ 最多可支持 32 路 TVOC 测量(配合 MD9908)。

⑦ 最多可支持 32 路粉尘浓度和 AQI 测量(配合 MD9912-PM)。

⑧ 最多可支持 256 路开关量输入(配合 MD9904)。

⑨ 最多可支持 32 路标准模拟量输入(配合 MD9911)。

设备外观如图 4 所示。

图 4　MD9102 外观

传感器接线如图 5 所示,在接入同一 MD9102 前,各传感器必须设置为不同地址。

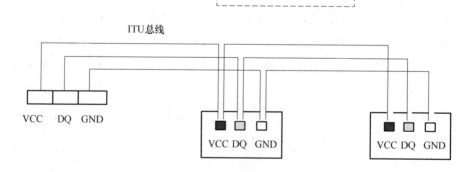

图 5　MD9102 挂接传感器示意图

MD9102 模块的工作流程分成两个部分:入网流程和数据上传流程。

① 入网流程:在此阶段,MD9102 模块会尝试加入 LoRaWAN 网络,每次用时 10～90 s 不等,此时 LCD 显示如图 6 所示。

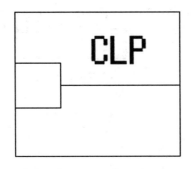

图 6　MD9102 上电屏幕显示图

若未能接入网络,LCD 显示界面如图 7 所示,E01 表示通信故障,E02 表示入网失败,模块会从头开始,重新进行入网步骤。

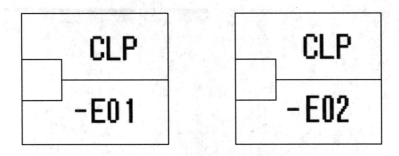

图 7　MD9102 上电后未能加入网络

若能加入网络,LCD 显示界面如图 8 所示。

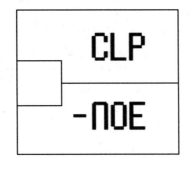

图 8　MD9102 上电后正常加入网络

② 数据上传流程:在此流程中,模块会定时采集所接 ITU 产品数据及开关量输入数据,并将数据上传至网络。LCD 显示界面如图 9 所示。

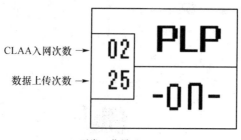

正常工作显示

图 9　MD9102 数据上传 LCD 显示图

2. MD9908

MD9908 是 TVOC(总挥发性有机物)传感器,能返回现场空气中 TVOC 的数据,适合于与其他环境监测 ITU 设备灵活组建现场监测应用系统。其外观如图 10 所示,技术指标如表 2 所示,多点联网接线使用如图 5 所示的方式即可。

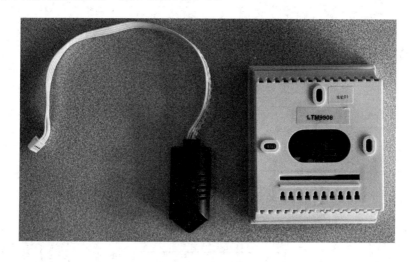

图 10　MD9908 外观

表 2　MD9908 技术指标表

型号		MD9908
结构	外形尺寸	100 mm×80 mm×25 mm
	结构形式	一体式(可定制分体式)
TVOC	测量范围	$0 \sim 17 \times 10^{-6}$
	分辨率	0.001×10^{-6}
供电电压		5VDC(支持 MD9662、MD9600 为其 POE 远程供电)

3. MD9912

MD8912 是 AQI 粉尘传感器,能一次返回 PM1、PM2.5、PM10 和 AQI 的数据,适合于与其他环境监测 ITU 设备灵活组建现场监测应用系统。其外观如图 11 所示,技术指标如表 3 所示,多点联网接线使用如图 5 所示的方式即可。

图 11 MD9912 外观

表 3 **MD9912 技术指标表**

型号			MD9912
结构		外形尺寸	98 mm×85 mm×45 mm
		结构形式	一体式
性能参数		测量原理	激光散射
		PM1	AD 0 通道 0～5 V(0～1 000 μg/m³),分辨率 1 μg/m³
		PM2.5	AD 1 通道 0～5 V(0～1 000 μg/m³),分辨率 1 μg/m³
		PM10	AD 2 通道 0～5 V(0～1 000 μg/m³),分辨率 1 μg/m³
		AQI	AD 3 通道 0～2.444 V(0～500 指数),分辨率 1
		工作电流	≤140 mA
		响应时间	≤90 s
		工作湿度	15%RH～80%RH(无凝结)
		工作温度	−20～40 ℃
		存储温度	−40～60 ℃

4. LTM9901

LTM9901 是便携式的温湿度传感器,其外观如图 12 所示,多点联网接线使用图 5 方式即可。

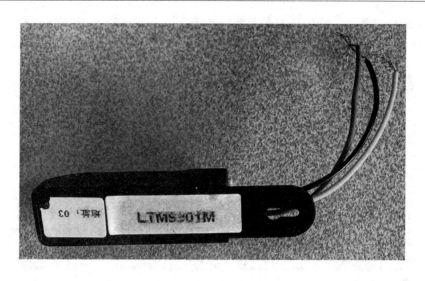

图 12　LTM9901 外观

5. MD9103 和 MD8877

MD9103 为 LoRa 多点数字温度传感器采集模块,可以挂接 MD8877 数字温度传感器,有 4 个传感器接入通道,有 2 路光电隔离型开关量输入,电池供电(外供电模式可选)。MD9103 定时启动数据采集,把现场多个温度测量点及开关量数据上传到 LoRa 网络上,可以组建多种大小不同的现场监测应用系统。

设备外观如图 13 所示。具体使用方式见 3.3.2 节 Experiment 2 实验原理介绍。

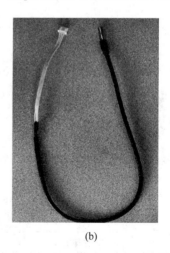

(a)　　　　　　　　　　　　(b)

图 13　MD9103 和 MD8877 外观